Massively Networked:

How the convergence of social media and
technology is changing your life

Pamela Lund

PLI Media, San Francisco

Published in the United States by
PLI Media, San Francisco
www.plimedia.com

Massively Networked:
How the convergence of social media and technology
is changing your life.
www.massivelynetworked.com

ISBN: 978-0-9831995-0-2

Interior Design by Joel Friedlander,
www.TheBookDesigner.com

Graphic Design By Tina Baumgartner

First Edition

For you
who do not know
everything is impossible

Table of Contents

The Massively Networked World

We are moving toward a world where interconnections among people, places and things are becoming increasingly tangible and more intricately connected. This massive network has no single center or central controlling force. You have more power to change the world than ever before. And so does everyone and everything else.

"If your computer could do anything you wanted," you ask ten-year-old Emma from Australia, "what would that be?"

Emma's eyes light up. "I want my computer to turn pictures of food into food I can eat!" She looks at you intently, as if she's placing an order she fully expects you will deliver.

It's not surprising at all that someone Emma's age would expect computers to be able to turn pictures of food into the real thing, especially when we consider how fast things have changed in her brief lifetime. Halfway around the world, eleven-year-old Jacob from Denmark wants a virtual computer—one without any hardware whatsoever: "I want a life-size 3D computer that I can walk into and it does what I want when I talk to it." Jacob's eight-year-old brother Daniel one-ups him with an even bolder command—Daniel requires his computer to be psychic: "I want a Helper Computer: it already knows what you want, and it does it for you." Thousands of miles away, nine-year-old Alex in Florida envisions his future as a virtual rock star: "I want a computer that's like Guitar Hero®, but with drums, too—and surround sound. And I want to play it with a touch screen."

Each of these kids looks forward to interacting with technology seamlessly. They fully expect the distinction between the real and the virtual to be fluid. And the kids are all right.

Latitude 42, a technology research company, conducted a survey of 146 children under twelve from around the world, asking them what they would like their computers to do. The answers were similar to those of Emma, Jacob, Daniel, and Alex. What you might *not* know is that of the kids' requests, more than 95% are either on the market right now or feasible in the next decade.[1] While time travel may still be a bit far out for human beings, physicists have observed teleportation at the quantum level. Who is to say when we may see a breakthrough on this front as well? Time travel aside, things are about to get very different very fast and the kids' expectations are a tiny preview of where we are headed.

For more than a decade, I've been watching digital technology trends and translating them into interactive and social media marketing plans for large businesses. Since the mid-90s, I've kept an eye on anything that can connect to the Internet—from people to mobile phones to satellites— along with the means of gathering and translating the mammoth amount of data being fed into the infosphere and the impact it has on our lives. Collecting and analyzing articles, videos, academic reports and blog posts allows me to spot trends early and figure out how they apply to my clients. In early 2008, I noticed several trends ramped up all at once: mass adoption of social media, infrastructure and bandwidth improvements, new user interfaces and mobile devices, immersive environments, the rise in games and their profitability, sensor network technology developments, search engine sophistication, artificial intelligence, analytics tools and robotics. I suspected that the next wave of change would not be simply adding more tech tools to the digital toolbox. More is not always just a relative measure of quantity—sometimes when trends converge they start to modify each other and create their own feedback loops. It occurred to me that this is one of those rare instances when 'more' changes *everything*.

The rise of social media and technology is at the core of this change. As of late 2010, an estimated 1.8 billion people are participating in the digital social network. Not only is the number of human participants climbing rapidly, but we humans are now being joined by billions of nonhuman agents. IMS Research estimates that by 2020 we will see upwards of 22 billion devices connected to the Internet and communicating online.[2] The interface among human beings, objects, and the infosphere is becoming intuitive, less cumbersome. The interaction between digital and physical reality is increasingly seamless, more immersive. Technologies for turning an inconceivably huge collection of data into useful information are evolving rapidly—they are more intelligent, more intricately interconnected. Each agent in the network—even nonhumans—can directly affect the network easier and faster than ever before.

The vast social kaleidoscope of people and things, combined with the ease of interaction *plus* the improving capacity to make sense of all this data, creates a massive network of feedback loops that dynamically inform each other in complex ways. By participating in this massively networked ecosystem, you as an individual have more power than ever before to change the world. And so does everyone and everything else.

Dozens of books exist that speak to digital and social media's impact on traditional business and how traditional businesses are being challenged by current socio-economic realities, so I will not devote space to that here. Of much more interest to me—and soon to you—is how the rise of social media and technology is changing your life in unprecedented ways. These changes give you more opportunity than ever to turn your ideas into reality—if you learn how to actively engage them.

Yes, those kids are already right on track.

Making sense of the world we're moving toward

How is one to make sense of all of this new input, anyway? This issue came up even while I was preparing to write this book.

With the explosion of information, my old way of organizing concepts and research in a relational database organized into a hierarchy of flat files became difficult to manage. The number of intricately interrelated categories and subcategories pushed me toward a more dynamic, less hierarchical information organization system. On page 6 is one view of the research map I created while writing this book.

The page six chart provides a great visual of the social media and technology elements that are beginning to converge; it is also an example of how a hierarchical, linear structure is insufficient to capture the complexity of the world we are headed toward. Though the technical category names and complex relationships among them may give you the impression that this book will be a hard read, it is not. My goal is to capture the essence of the world we are quickly moving toward and share that with you in nontechnical, jargon-free language. The research map is one of the tools that helps me to do so.

After twelve hours of setting up a new, nonlinear database, I could clearly visualize the convergence of technologies that I had started to notice. Selecting any of the categories, like "Internet of Things," brings that category and all its related research front and center making it easy to see how trends interrelate, and to create new relationships and delete relationships that turn out to be tenuous. Selecting a topic view I can see a bar chart showing which trends have the most research (articles, blog posts, videos, academic data) attached to them, giving me a quick snapshot of which trends are generating the most conversation. Go to www. webbrain.com to see live examples showing how this nonlinear database works.

How this book is organized

Writing a book about the massively networked world is like using a telegraph to describe radio. The written word is a mechanism for reducing complexity into linear and logical thoughts: it is, therefore, a limited medium for a message

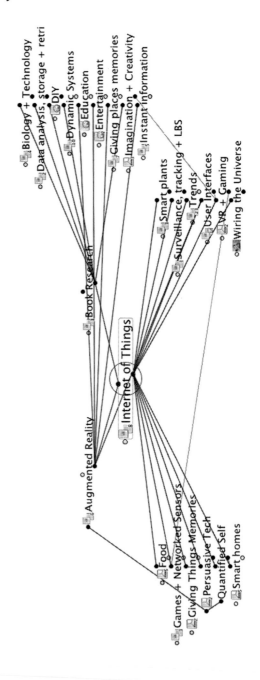

meant to describe a dynamic, nonlinear and fundamentally creative process. To get around some of the limitations, I have included Quick Response (QR) codes like the one below for an added dimension to the written content. You can download a QR code reader to your mobile device and point it toward the code. The code will take you to a video or resource relevant to the content of that page. QR code readers are available from a number of providers. Go to www.mobile-barcodes. com/qr-code-software for a list. Your purchase of the print version also gives you access to the online version. The online version includes full graphics and embedded links and videos not directly accessible in the print version. To view the online version, go to www.Massively Networked.com and select the book link. You will be prompted to enter the user name "mnbook" and the password found inside the back cover of the dust jacket.

Mobile
Bar Codes

The chapters that follow tell the story in detail, using real-world scenarios that exemplify how the rise of social media and technology is opening up opportunities for you to change your life—including how you work and think of business, how you shop, your home, the transformation of education and entertainment, changes in health care and prospects for longevity, civic engagement and politics, and finally, how we can best participate in shaping the ecosystem of which we are a part. It is worth noting that the real-world scenarios are, in the end, convenient thought-starters based on what is happening today. They do not necessarily prescribe a vision for the future. Whatever that future may be is becoming increasingly unknowable as the pace of change accelerates. Ultimately, this book is more about how best to engage this world of accelerating change than it is about the changes themselves. Before moving into concrete examples, it is key to understand the role of imagination.

CHAPTER 1

Your Creative Reality

Imagination is as practical and important a part of the rational process as logic. Through imagination, you have access to a vast store of immaterial realities that have the potential to shape the physical world.

At a recent South by Southwest Interactive conference, I attended a presentation on science fiction's impact on reality. The speaker said she has been "blown away by how well science fiction authors predicted so much of what future technology might look like." You can understand why the speaker would think certain science fiction authors were prescient enough to pick up on what might happen forty or fifty years down the road—especially when you can recognize technology from *Star Trek,* like Captain Kirk's communicator, in your hand today. That science fiction writers had particularly keen foresight is not, however, the connection between sci-fi of the past and technology of the present. The future is not on a pre-determined trajectory that simply has to unfold. Putting to the side for a moment that predictions of the future are notoriously inaccurate, *there is a direct re-*

lationship between what we imagine and what ends up being real.

On a basic level, stories and experiences that resonate with you shape your choices and your behavior. It is no accident that you live in a world where airplanes, rockets, robots and virtual immersive environments exist. Stories of human flight have been told and retold since Daedalus gave Icarus wings. Robots and human replicants entered the imagination with the rise of the Industrial Age, and Fritz Lang's *Metropolis* brought robots to the movies in 1927. Stories of virtual realities have been around since Plato's cave and in tangible form since Martin Heilig's invention of the Sensorama in 1960. While science fiction writers do not predict the future, *they very well may have a hand in creating it.* This is a more important idea than you might at first think.

As the potential to affect change enabled by technology and social media ramps up, understanding the practical role the faculty of imagination can play will be key to grasping its increasingly important place in our lives. At present, imagination is not very well understood, nor is it taken seriously aside from pop psychology or metaphysical books that advocate creating vision boards or thinking your way to riches. Imagination is typically given its due in the realm of art, abstraction, and dreams. It is not, however, usually considered the practical faculty that it is.

Fundamentally, imagination is an act of creation. Such creation may be as simple as a thought or short story, or as complex as an entirely novel framework for understanding the world. Whether that creation becomes tangible or remains imaginary depends on a complex variety of factors including timing, resources, known physics, resonance with others, and level of commitment. The mechanism for turning imagination into reality is usually not as simple as checking that vision board off the list and waiting around for it to manifest. What we end up making real is frequently influenced by imaginative acts of creation that went before and the stories and experiences they left in their wake. We are swimming in a sea of stories—many of which we are barely conscious. The bulk of this book is devoted to shining a light on stories we are all living with and likely take for granted, with the goal of opening a door to creating better ones. Imagination is key to this process.

Imagination as practical faculty

Albert Einstein is famous for his astute quotes about the creative process—and it is probably safe to say that he should know of what he speaks. "Imagination," he declares, "is more important than knowledge." He also states, "Logic will get you from A to B. Imagination will take you everywhere."

You could take Einstein's quotes to mean it is good to daydream as an escape from the real world. However, Einstein is likely saying something much more profound than advocating escapist fantasizing. If he were alive today, he likely would agree with me that imagination is not only the mother of dreams and art but is also fundamental to creating new paradigms or novel frameworks for organizing our perception of the world. Ideally the new paradigm or story makes sense of the world in a way that is more elegant or satisfying than the one it is replacing.[3]

Let's take Einstein's Theory of General Relativity for a concrete example of how the imaginative faculty can work. The Theory of General Relativity fundamentally altered how theoretical physicists began to think of the dynamics of time, space, and their relation to the "force" of gravity. (According to Einstein gravity was not a force at all, but rather a result of the geometry of space-time itself). Einstein's theory is a framework derived from an instant of imaginative perception of universal phenomena rather than one built by simply adding to or altering existing theoretical paradigms. He describes this instant of realization as the result of an unconscious process: "A new idea comes suddenly and in a rather intuitive way. That means it is not reached by conscious logical conclusions. But, thinking it through afterwards, you can always discover the reasons that have led you unconsciously to your guess and you will find a logical way to justify it."

Einstein believed his intuition was "nothing but the outcome of earlier intellectual experience"—experiences that had presumably aggregated in his mind, floating like pieces of a puzzle just waiting for the opportunity to snap into a proper framework. While his interpretation begs the question of what it is that made this framework so robust, it does give credence to the idea that he believed imagination played a practical role in creating the framework within which the fabric of space-time might be understood. Just as Copernicus' heliocentric vision of the universe eliminated the need for the complex workarounds to predict planetary orbits required by Ptolomy's geocentric model, Einstein's theory resolved puzzles inherent in previous physics theories and provided a more elegant way to make sense of universal forces.

Calling imagination a "faculty" that makes sense of some type of input—like sight, hearing, taste, touch or even the mind—might seem unusual. If the ears respond to vibrations of various frequencies traveling through the air and translate them into sound, what is it that the imaginative faculty responds to?

There are three possible answers to the question of what "input" the imaginative faculty responds to. They can be distinguished into modes of imagination: inductive, fantasy and creative. These three are all relevant for making sense of the world we are moving toward.

Inductive imagination

Inductive imagination gathers input from ordinary experience and puts these elements of experience together into a fresh interpretation. This mode has a lot in common with inductive reasoning in that it takes known ideas or data and extracts new meaning by interpreting how different things might "hang together." Similarly, inductive imagination brings something new into the world through reinterpretation of things that already exist.

The product of inductive imagination is closely aligned to the theme and variation we find in the fashion industry. Each season draws from a vast library of fashion components including colors, fabrics, shapes, and functions to produce a line that is at once familiar and novel. Its novelty comes from putting together elements from the archives in ways that have not been seen in seasons past. The elements themselves, however, do not typically give rise to an entirely new paradigm for understanding what "fashion" is.

Musical imagination is an inductive mode similar to that of fashion in that it also relies on a vast library of elements including sounds, instruments, tempos, and chords that can be combined to produce new music. Stewart Copeland, drummer for The Police, explained in a *Studio 360* interview that The Police's musical strategy was to take familiar reggae beats "and move the rhythm one beat to the right" to come up with a new sound for their song "Roxanne."[4] This is classic inductive imagination at work.

Inductive imagination is perhaps the easiest imaginative faculty to gain some mastery of, and one that I use frequently as I work with clients. My process requires identifying the raw components I have to work with and the end result the client would like to achieve—whether it be generating new leads, improving brand perception, or increasing sales. The resulting strategy brings together the raw components to help clients meet their business goals in a way that is ideally more effective than whatever strategy they had employed before.

For example, a client who has an outdated brand may ask me to come up with an online strategy for updating brand perception in a way that resonates with an audience likely to purchase their product. The raw materials I have to work with include their product, current brand perception, potential target markets, the budget and resources available for the rebranding effort, and sales goals. Say the product is jam. The current brand perception is this company makes the same predictable flavors of jam that they have made for decades. While that is not necessarily a bad brand perception to have, it will not work toward meeting their goal of growing the company's market share of jam products. My role is to identify a new target market for a jam product and imagine what type of jam might appeal to them. I notice, for instance, that in other food product categories, crazy flavors are the rage. I might suggest creating a pilot product— say, mango-lime jam. I then create an interactive marketing strategy that will appeal to GenXers who are currently purchasing other foods in this crazy flavor category. A great strategy starts with a fully imagined landscape that includes mango-lime jam, GenXer personas, current events, the time of year—and the list goes on. Creating a strategy this way uses the inductive imaginative process.

Fantasy imagination

The next imaginative mode is fantasy. Through fantasy imagination, you can extend elements from common experience in novel and unexpected ways. This is the kind of imagination that drives much science fiction, horror, magical realism and other genres of entertainment that are at once familiar and strange. Fantasy brings something truly new into the world by twisting everyday reality into unexpected shapes that allow you to see the familiar from an alternate perspective. *World of Warcraft* is a fully imagined fantasy gaming world where multiple players are embodied in familiar yet strange virtual avatars that vie for power according to familiar yet strange rules of engagement. Participation requires a suspension of "real" world notions of form and interaction while engaging the alternative reality of the fantasy world. This sort of out-of-body experience opens a gap in the fabric of reality that can let new insights emerge.

The early *Star Trek* series engaged the fantasy imagination of its viewers in a similar way. The series subtly opened a gap in the fabric of 1960's reality to generate insights about everything from race relations to environmental conservation. The psychological distance created by framing the story in a futuristic sci-fi setting allowed many viewers to become comfortable in a world where Uhura, a black woman, played an important role in running the Starship

Enterprise without overtly disturbing mainstream 1960's prejudices about race and gender.

Creative imagination

The final imaginative mode is creative imagination. This mode is a kind of hyper-conscious state that lies between dreaming and being awake to the mundane world. In its fullest realization, it is a state where the perception of time disappears, where "thinking" is not possible without exiting the mode of creative imagination, and where it seems everything that can exist does exist as an immaterial yet fully real whole. It is the realm described by so-called mystics—and a fair share of madmen, too. It is a rare state of perception and one that will literally blow your mind if you let it. Carl Jung's description in his autobiography *Memories, Dreams and Reflections* of intentionally descending into a kind of madness to plumb the depths of what he called the unconscious hints at the mind-blowing potential of this state when fully engaged.

On a more mundane level, creative imagination is the process of apprehending something original that is ready to be born into the world in some form but does not yet exist—for better or for worse. I would put Einstein's General Theory of Relativity and the mechanical art of Leonardo da Vinci in the category of imaginative insights that created something truly novel in the world. Other creative works may also belong in this category such as Frank Gehry's Guggenheim in Bilbao, Spain, which arguably constitutes an utterly new form of architecture. Contrast the Guggenheim with the Venetian Hotel in Las Vegas, the latter of which is the a classic result of fantasy imagination that takes an existing form and twists it into something at once familiar and strange. While the bulk of fashion and music primarily belongs in the inductive category, entirely novel fashion and musical leaps could be the result of this kind of creative imagination as well.

Through the imagination you have access to a vast store of realities that have the potential of shaping your physical world. These realities can be stories from your experience and the aggregate experience of others. Or, at the most fully engaged creative mode, they can be realities that generate something entirely new in the world. The faculty of imagination does take practice to develop, especially since it has been relegated to the sidelines for so long. Just like the nose of a wine tester can be trained to distinguish subtleties of smell to identify which components tend to make the tastiest pinot noir, or the ear can be tuned in to hear nuances in music to identify which elements tend to create the most inspiring sonatas, the imagination can be tuned to conscious awareness. Why is engaging the imagination especially valuable now?

Engaging a perpetually changing reality

As the pace of change ramps up, relying on stagnant stories for organizing our lives becomes more cumbersome. Many of us are starting to realize that in order to live a fulfilling and satisfying life, an increasing number of work-arounds are necessary—including complex strategies for managing finances and finding a way to push aside activities we would find more fulfilling in order to remain employed. The daily grind of spending time focused on tasks like grocery shopping, figuring out meals, coordinating schedules and maintaining the home seems to fill most of the rest of the time. In the end, these activities are driven by stories that say we must work hard to earn a paycheck, that material possessions are valuable in themselves, and that time and joy must be sacrificed for later gratification. Engaging in a perpetually changing reality successfully requires an entirely different strategy than relying on a static set of stories—it requires actively engaging the imagination. Imagination is needed now more than ever if we want to create a more satisfying vision for how we want to enjoy our lives, especially as opportunities to do so accelerate.

While all three imaginative modes just described—inductive, fantasy and creative—can play a part, it is creative imagination that stands the best chance of dynamically producing new frameworks for reality to take the place of the one that is reaching the limits of sustainability. Unlike any other time in history, however, this creative output need not be from a single source or a small group of visionaries with the means to communicate their vision. We now can use the convergence of social media and technology to harness the creativity of a multitude of people.

Unlike Einstein's "puzzle pieces" that he gathered over the long trajectory of his own intellectual experience, the source of creative genius now can be collective and virtually instantaneous. The puzzle pieces brought to reality by the convergence of social media and technology can pop into life from the power of multiple brains and bodies all over the globe and almost at once. The only requirements are desire and intention to participate. The new paradigms that emerge will be the ones that most closely align with the intentions of its participants.

The stories we tell ourselves matter

Just as frightening military robots straight out of *The War of the Worlds* exist today alongside helpful service robots straight out of *The Jetsons*, it matters a great deal which stories you choose to tell and which you choose to take in. And ultimately, it is your choice and responsibility. This is one important reason it matters more than ever to become conscious of the kind of world you imagine. The world you imagine is the creative ordering principle that drives your desires and intentions, just as the worlds others imagine drive theirs.

*Boston Dynamics'
"Big Dog" video*

*Willow Garage's
service robot video*

Another reason it matters more than ever to become conscious of the world you imagine is the potential for collective acceleration of change. Combining the collective power of imagination can accelerate the process of creating new frameworks that are just right for the world we want to live in; as a result, we can more quickly leave ordering narratives that no longer work very well behind.

What was said in the Introduction bears repeating: you as an individual have more power than ever before to change the world. And so does everyone and everything else. Therein lies what is different. What we imagine individually and collectively *matters*, quite literally, now more than ever before.

CHAPTER 2

The Massively Networked "Toolbox"

The tools that connect you with others and turn your ideas into reality are increasingly powerful. The only obstacle to creating the world you want is a failure of imagination.

At heart, I am a "presentist." I am interested in the dynamics that shape our reality as opposed to a futurist who speculates on what that reality might be. I am neither a techno-utopian who believes we are improved by being implicated into "the machine," nor am I an eco-utopian who finds the greatest wisdom in nature. Nature and technology are part of one and the same ecosystem, and the interconnections are becoming more fluid as technology is beginning to rival nature in its intelligence.

Your perspective as a human being has the chance of being expanded in increasingly tangible ways as you participate in and contribute to this ecosystem, if you pay attention to what is happening around you and learn how to make the best use of the tools and your talents. The tools now becoming available for your use are formidable. They give you increasing power to connect with others, to accelerate the growth of your intelligence and to shape the world you live in. With such tools at your disposal, the greatest obstacle to creating the world you want will only be a failure of imagination.

Here is a sampling of elements in your new toolbox:

Social Media

Social media is a fundamentally dynamic arena that encompasses both social networks, like the people who choose to interact on Facebook or LinkedIn today, and the technology that connects them—everything from the physical "pipes" that transmit data, to the bits and bytes of data itself, to user interfaces and software that organize which pieces of the network are connected to one another. It is worth highlighting that you cannot reduce social media to either its content (such as blogs, feeds, and posts) or platforms (like Facebook, Twitter, LinkedIn). Both content and platform are simply artifacts that have little meaning outside of the dynamic, interactive context of the medium of which they are part. The dynamic interaction is what differentiates social media from, say, television or radio.

The difference between social media today and pre-Internet social media is essentially the difference between mailing a letter and telepathy. Mailing a letter is a slow means of communication. The farther away the members of the network are, the costlier the communication. There is a long lag between send and receive. It is virtually impossible to have more than one conversation at a time. Writing the letter, putting it into an envelope, stamping it, putting it in the mail, waiting for delivery and the other person to write a letter back, and so on, creates a lot of friction in the feedback loop.

Social media today eliminates a lot of this friction. You can have multiple conversations at once, each inter-modifying the other. The cost of communication is low. It does not matter as much where others are located or if you are available to have the conversation at a given time. Send and receive are virtually instantaneous.

As devices mediating communication become smaller and less intrusive, the closer communication approaches telepathy. And, as relationships among data in social networks are mapped, patterns of connection and interaction are made visible over time. All of this new intelligence can be used to predict activity, from your buying behavior and your health to chances that a particular business endeavor will succeed.

Social networks are now beginning to include nonhuman communicants, adding even more data to the social media mix. This network of nonhuman communicants is known as the Internet of Things.

Internet of Things

The Internet of Things conventionally refers to objects with sensors attached that are connected to the Internet. Objects could include the thermostat, appliances, and plants in your home that you can monitor and access remotely to adjust the temperature, turn off appliances, or water your plants while you are on vacation or at work. They could also be bridges and roads with sensors built in that tell your local traffic controller when to suggest alternate routes due to heavy traffic in a particular location. Because they are connected to the Internet, they can be monitored online and networked with other objects.

The networking of mundane objects is only one aspect of the world that is being networked. Projects are also actively in the works to wire cities, the planet, and even other planets and orbiting satellites into a massive global and interplanetary network. Just as things can be included in the social network, people can be included in the Internet of Things. The entire ecosystem of which you are a part is being tangibly interconnected into something resembling a universal central nervous system—with fascinating implications.

Augmented Reality

Augmented reality (AR) places a layer of virtual information over a physical setting. AR can recognize a face, brand or location when you point your smartphone in that direction. Once recognized, an overlay of information floats in the foreground of your screen while the physical object you are pointing toward remains in the background. For instance, if you are at the physical location of the U.S. Civil War's Battle of Gettysburg, you can point your smartphone to the site and see an information overlay letting you know who gave a speech on that spot while the current landscape remains in view in the background. AR can also place virtual objects in a physical setting. Pull up an AR-enabled furniture catalog on your smartphone, select a couch, and you can immediately see how it would look in your living room.

The Quantified Self

Persistent monitoring and recording of your entire life is now possible. You can set up an account to live stream and archive as much of your day as you like through your mobile phone. You can have your physical activity monitored and automatically fed to the Internet. Your weight, sleep patterns, heart rate, pulse and almost any other biometric you can think of can be monitored, streamed online, and archived. Every eBook you read, everywhere you go, and everyone you visit can become part of your permanent record. You are able to map every thought you have to an online "brain." This online brain might be a visual relational database like the one found at TheBrain.com, where you manually enter every thought you choose to record. All of this monitoring can reveal patterns and habits you are likely not aware of and may want to improve or eliminate. The goal of the quantified self can be as modest as identifying opportunities for optimizing your health, or as ambitious as creating an exhaustively comprehensive virtual version of yourself that may later be ported into a life-like virtual or robot avatar.

Robots!

Robots include everything from nanobots that can manipulate matter at the molecular and even atomic level, to functional robots optimized to perform a specific task like cleaning your floor or disarming a bomb, to humanoid robots that may be remote controlled allowing you to attend a conference and interact with attendees, to autonomous robots that perform services. Robots are not, strictly speaking, part of the social network—yet. They are, however, getting smarter and more functional and will play some part in the stories to come.

Location-based services, surveillance and tracking

Location-based services (LBS) use the global positioning system (GPS) to identify where you are on earth and provide some kind of service. Services like Facebook, Foursquare and Gowalla let you check in to places and tell your followers where you are. Rack up enough Foursquare visits at a local restaurant, and you may win a title like "Mayor of Bistro Jacques." You can also see how many other people are at the same location. This can be very helpful during a conference if you want to head to where the party is or, conversely, find a quiet place for dinner.

Giving a service permission to track your movements in exchange for their services is one kind of tracking. Surveillance is another. Surveillance technologies have become good enough to recognize faces and make intelligent decisions as to what from among the persistent stream of data is worth alerting security about.

Tracking behavior online is ubiquitous. Online advertising networks that use re-messaging or retargeting know when you have visited a site and serve up ads from that site wherever you go within their network. If you were wondering why Overstock.com keeps pitching you that duvet cover wherever you go online, it is not because Overstock has paid for a huge number of ad placements. They are serving up ads to you in particular because they know you have looked at the duvet covers on their site; they are betting that interaction means you may be interested in purchasing one from them—if you are reminded enough times.

Persuasive technology

Persuasive technology rests on the premise that technology plus social networks can change your behavior. In an old-fashioned way, this is how a service like Weight Watchers has become so successful. Weight Watchers uses a point system that allows you to eat what you like as long as you do not exceed your point quota for the day. The point system makes you more aware of choices and tradeoffs when it comes to eating. You can choose to eat those French fries for lunch, but you might have to skip dinner that night. You are also encouraged to participate in check-ins either online or in person. The accountability of participating in the social network dramatically increases your likelihood of successful weight loss.

Persuasive technology takes the premise of accountability and awareness to a new level. By bringing individuals or organizations into a social network and providing a way to make their choices and the tradeoffs conscious, persuasive technologies can directly modify your behavior. OPOWER (www.opower.com) is a software company that uses a simple version of persuasive technology to help homes become more energy efficient. Subscribers can immediately get a read on their energy consumption compared with their neighbors' through a home display unit, online report or smartphone. Those in the neighborhood who use relatively less energy see a smiley face by their readout. OPOWER says this causes 60 to 80 percent of people to modify their energy usage.

Data storage and processing

Data storage media are getting exponentially smaller as the amount of information it is possible to store grows exponentially larger. All the while, the devices we use to access all this information shrink in size and the whole lot grows less expensive. Moreover, enormous data storage systems can be used by multiple entities in the "cloud." The difference between traditional computing—where you would have all your computing needs taken care of by a server that you own and control—and "cloud" computing is the difference between having your own generator to power your home versus having your home connected to the electrical grid. By keeping computing services in the cloud, you can leverage network efficiencies to provide storage and processing much less expensively than individual standalone systems. In short, because of the exponential increase in efficiency of material production, storage and retrieval processes, powering and paying for increased data storage and processing needs should continue to become less of a barrier.

Data analysis and retrieval

Data itself is meaningless. Only when data is connected does it turn into information: the more relevant the connections among pieces of data, the more useful the information.

I worked on development of a search engine (Snap.com, acquired by NBCi in 1999) before Google arrived on the search scene. As you would expect, our team's job was to ensure the most relevant results surfaced when someone performed a search. We quickly realized the limits of keyword matching and assigning relevance based on where a word or phrase appeared on a web page. Each day I could see trending search terms emerge that could be related to any number of web pages with the word or phrase appearing in heavily weighted sections of the page. For example, we observed in the case of the search word "air" that the increased number of searches was due to the electronica band Air getting a lot of radio time (yes, radio). Data analysts joined the team to build rules for surfacing results based on inferred connections between data points. In the "air" example, searchers who selected the band link from among the other links in search results might share the characteristic of, say, liking electronica music. The search engine *learned* that searchers with similar preferences would likely be looking for the band Air, and not a site about air quality or some other attribute associated with the word. By using rules of inference based on clustering of terms in online profile data, the search engine could serve up a more accurate result to you, the individual searcher.

In the last ten years, the amount of data being fed into the infosphere has increased exponentially along with the capacity for making sense of all this data. To say the infosphere is *learning* at an increasingly rapid pace is not a metaphor. With the rise of semantic analysis, mapping of the social graph, sentiment analysis and rapid improvements in artificial intelligence, accurately connecting the data and turning it into useful information is taking on a life of its own. Couple this with the ease of people and things contributing and retrieving data through online interfaces and mobile devices, and you build a massive network of feedback loops that go way beyond linear information processing and input/output. In a very real sense the infosphere is an entity unto itself in that it holds memories, makes sense of what it takes in and shares information with others. The infosphere is an entity we participate in creating and are simultaneously affected by—much like a river both affects and is affected by the riverbed through which it flows.

Human-computer interfaces

A human-computer interface (HCI) is any device that makes communication between you and smart machines possible. A smartphone is one kind of human-computer interface. With a few touches, the phone connects you to the Internet. The ideal HCI is one where the connecting device becomes invisible and the interaction between you and the machine approaches instantaneousness.

Working HCIs currently include brain implants that allow a paraplegic to work on a computer and control his wheel chair, skull caps that capture and translate brain electrical activity so you can remote control your mobile phone by thinking commands, and contact lenses embedded with devices that monitor biological functions and wirelessly transmit data to a receiver. As HCI technology becomes more easily accepted by and integrated with biological systems, ease of interaction with machines will begin to blur the boundaries between where you start and where technology begins.

Gestural and haptic user interfaces

Gestural user interfaces (UI) use your body movement to manipulate data and objects in your environment. The movie *Minority Report* is one of the most well known visualizations of how gestural UI can work. In the film, Tom Cruise's character flips through virtual data screens projected in front of him using only the movement of his hands. Today, working prototypes allow users to take a picture by simply creating a frame with the thumb and forefinger of both hands, or to resize pictures projected in front of them using a pinch or reverse-pinch finger motion.

Coming along right behind gestural UIs, which use physical movement, are haptic user interfaces that engage your sense of touch to convey information or an experience. Developments are in the works to add haptic information to the gestural UI as well, so if you want to move a file to the trash using a gesture, more important files will feel heavier to move, giving you time to deliberate, while less important ones will be easier to trash. Virtual reality games and immersive environments are adding haptic information to give you the sensations relevant to the environment you are experiencing. If you are taking off in a rocket ship in a virtual reality experience, for instance, you might feel the pressure of the g-force or the vibrations of lift-off.

Virtual reality, 3D environments, and gaming

The promises of virtual reality and 3D immersive environments have captured the popular imagination since the *Star Trek* holodeck appeared on our television screens. Captain Kirk could walk onto the holodeck, close the door, and time travel to 1940's Chicago or be entertained by a sexy alien chick. The difference between real and virtual was portrayed as almost imperceptible in the holodeck environment. Improvements in 3D technology coupled with voice command recognition and enhancements in gestural and haptic UIs now allow you to project yourself into a virtual environment and feel what it is like. Xbox Kinect is one example of the beginnings of such an immersive virtual experience currently available.

Alternate reality games engage multiple players to deal with what is frequently a real-world scenario like running out of oil. These immersive experiences help you enter a mindset as if the scenario is real and allow you to devise solutions to challenges that arise and see how they play out along with the solutions your co-players have tried. The massively multiplayer experience means that your actions in the game affect the ecosystem of the game as a whole, creating outcomes that could not be arrived at through a linear, logical process.

Entering virtual reality and immersive environments is not confined to indoor activity only. Augmented reality applications projecting alterations on the physical world that you can interact with using gestural UIs and feel through wearable haptic devices turn the world into a virtual/real hybrid, which has all sorts of possibilities for both fun and practical applications. Say you are deathly afraid of spiders. At home you know you can control your environment to a great degree, but outside you have little to no control. Through an AR application, you can engage in exposure therapy outdoors. As virtual spiders pop up out of nowhere and you feel them crawling on your skin, you have a chance of desensitizing yourself and getting over your spider phobia.

There is a rising interest in implicating game dynamics into social media to influence behavior. Game theory, along with persuasive technologies and the quantified-self movement mentioned before, all converge around the goals of identifying, analyzing and shaping patterns of behavior. Gaming is a dynamic arena where these goals can be realized almost real time. *World Without Oil*, a massively multiplayer alternate reality game, proposed a scenario where the world was set to run out of oil in 32 days. Eighteen-thousand participants from twelve countries engaged in the game as if the

World without Oil
video

scenario were actually happening. The game harnessed the collective imaginations of participants with the goal of dynamically generating creative responses to a world without oil. You may have always seen life as a game. Now you have more tools and intelligence at your disposal to play it better.

DIY

Do-it-yourself (DIY) culture is in evidence across disciplines from publishing to health care to science to manufacturing and production. Never before have the tools to write, publish, distribute, and market your own book or piece of music been so inexpensive or accessible. The depth of available information on any topic you are interested in makes self-education on technical disciplines like science and medicine more possible than ever. Prices of 3D printers that can take a computer-generated diagram and print out a working machine or copy a physical model are decreasing as their functionality is getting better. There are even 3D printers that print out living cells that may have the potential to grow into working organs. The tools of all but the most sophisticated trades are more within reach than ever before in history. Like the reduction in communication friction, DIY reduces production friction by eliminating intermediaries and reducing the costs of trial and error. Turning thoughts into things is becoming an increasingly easier and faster process.

A note about real world stories

The people and stories you encounter throughout this book may seem sci-fi at times, but they all use elements from this toolbox that are either in market now or have working prototypes, with minor exceptions. More importantly, the acceleration of social media and technology gives you more and faster ways of shaping and being shaped by the world you imagine. The feedback loops are speeding up, the barriers to material production are eroding, and we have more ways than ever before to personally and collectively make a difference.

Moreover, the real world stories that follow can provide foundations for new stories that we will create. Some stories will live a brief life and fade from memory quickly; some will have staying power and end up shaping our world. The stories are sparks to light the imaginative fire rather than prescriptions for the future. Consciously imagining the world you want to create is more important now than ever before.

CHAPTER 3

Your Massively Networked Self

You now have the opportunity to begin creating a quality of life that reaches beyond the constraints of what you may currently perceive as your body's physical limitations.

Imagine that you are working out at the gym one evening. Your weight, exercise routine, and body's response are transmitted wirelessly to the Internet. As you head for the locker room, your workout app automatically sends a tweet announcing that you've completed 35 minutes on the elliptical with an average heart rate of 60 beats per minute and you currently weigh 190 pounds. The same information is archived on your remote server with a time and date stamp.

Your partner sees the tweet, responds with a direct message asking where you would like for dinner. You start to reply "pizza" when the app pops up a message: "2 slices pizza est avg 580 cal = 1.25 hrs," reminding you how long it will take to burn off those calories on the elliptical based on your average workout intensity. You end up tweeting "salad" to your partner instead. "Could you pick up a couple of Mediterranean Salads up from the Café on your way home?" she asks you. Getting in the car, you tell the GPS "Café Organic." You are tempted to take a route you think is quicker, but experience tells you that ignoring the GPS will likely leave you stuck in traffic. You follow the GPS instructions even though you are unfamiliar with this route, and quickly arrive at the café.

Your genetic information has been analyzed, indexed, and stored with a service along with your health stats, real-time activity, and consumption data. You tweet a snapshot of your mediterranean salad to the service. Image recognition identifies its ingredients and automatically estimates the calories, nutrition, and ratio of fat, protein, and carbohydrates. You can adjust the estimates based on actual values for meals you eat frequently to improve accuracy of how the system as a whole estimates calories and nutrition for yourself and others.

A pedometer transmits how much you move versus sit throughout each day. A sleep pattern monitor transmits how long you sleep and how much of that sleep time is spent in the REM stage. The service cross-references all this data with comparable data others provide: it recognizes optimal patterns and suggests how you might optimize your daily wake time and sleeping routines.

After a few years of collecting and archiving data you learn that technology has become available to monitor your health from the inside. You wear contact lenses that monitor components of your eye fluid and body chemistry. The lenses have a chip that transmits status and any anomalies wirelessly to the server for further analysis. You can now get an injection of nanobots that circulate through your body identifying cells that seem to be multiplying more quickly than average to provide an early cancer warning.

You discover that despite your current commitment to health and fitness, your less-healthy past has had an irreversible impact. You will need a new liver. The long wait lists are a thing of the past. Your doctor orders a new liver to be grown from stem cells and within a few months you are back on the road toward optimum health.

In the intervening years, genes for longevity have been identified and researchers have figured out how to use these to reverse the aging process. You now have the option to live a much longer, healthier life. Choosing how long you would like to live is becoming a question you may have to answer in the near future. You now look and feel great, but will you reach a point where even that gets old? The answer is probably very different for different people.

Building self-awareness

If you are like most people, you do not think about the repetitive patterns that comprise your daily life. Such patterns are made up of mundane habits like brushing your teeth and driving to work that do not seem to warrant much consideration. They are also aspects of life that reside outside of normal awareness, like blood pressure and sleep cycles. Religiously monitoring and archiving data about food intake, exercise routines, sleep and activity patterns currently appeal to a small number of pioneers. The available self-monitoring tools still require time and attention to get the most out of them, and with relatively few participants at present, the data sets available to analyze are limited.

Gary Wolf and Kevin Kelly, both *Wired* alumni, coordinate a group attracting early adopters interested in building systems to improve self-awareness. Their blog at QuantifiedSelf.com covers topics from mood regulation, sleep optimization, thought mapping, nutrition, energy consumption, genetic testing, brain enhancement supplements, and monitoring gadgets, to tools and standards for making sense of all this data. Participants have personal goals from simple health optimization to creating a living portrait of themselves that can be ported into a robotic replicant or a virtual self housed online.

Bina48 video

Bina Rothblatt has a robotic avatar named Bina48 whose repertoire of knowledge includes everything Rothblatt has mapped and uploaded to Bina48's system.

Wolf says that while the tools and tactics for self-monitoring get the most attention, for him "the part that is the most interesting is the 'self' part. From the beginning what we were trying to investigate is how self-quantification generates changes in our inner experience." Much of what we believe to be true, he acknowledges, is influenced by everything from media messages to beliefs passed down through the generations. Collecting data about our behaviors can shine a light on aspects of our inner experience, such as unquestioned beliefs, which may turn out to be less accurate than we thought.

"Here's something that I just learned at a recent Quantified Self meeting," Wolf elaborates, "one guy created a metafilter called Fuelly (www.fuelly.com) that helps people track their gas mileage and how they drive. After collecting driving behavior data from fifty thousand people, he learned that even by becoming massively aware of your driving behavior and changing it to drive more efficiently, it actually saves trivial amounts of money. There is another guy who built an energy meter that included tracking thermostat use. Like with Fuelly, the amounts of money saved by adjusting the thermostat weren't high enough to warrant the monitoring. In the end, freezing or driving differently may *feel* morally correct, but it actually isn't a very powerful way to save energy."[5] It turns out that, in addition to optimizing behaviors, extensive monitoring may also help identify which behaviors aren't worth worrying about at all.

Adoption of life monitoring systems among the masses depends on demonstrating their value and making them easy to use. Much of the technology already exists—it just takes some effort to pull all the pieces together into a useful package at the moment. Once a simple to use bio-monitoring and feedback system demonstrates success in helping people lose weight, for instance, it is only a matter of time before the value of more extensive monitoring and modification gains popularity. These technologies are already beginning to gain traction.

Personal bio-monitoring technology

Monitoring sleep cycles is one bio-monitoring technology currently available. The Sleep Cycle application, for one, uses your iPhone's accelerometer to sense and track your movements during the night. The application correlates movements to where you are in the sleep cycle—light sleep, REM stage and deep sleep. If you set the alarm, the application will wake you when you hit the lightest sleep phase within 30 minutes of your alarm time. This is to avoid waking you during the deeper stages when the alarm would be more disturbing and you would awake feeling less rested. The application learns your patterns over a few nights so it can select the optimal time to wake you. It is not a big leap from monitoring your patterns to live-transmitting them online and cross-referencing that data with your other daily activity and consumption habits. You may learn, for instance, that eating a spicy meatball sub at 8 p.m. leads to a longer dreaming stage and less deep sleep than, say, the salad eaten at 6 p.m. Or perhaps you can finally make the connection that drinking half a bottle of pinot grigio in the evening usually means you will wake up four hours after going to bed—not a good strategy if you need to be awake and alert the next day.

Over time, the result is that you will be able to identify how your waking life behavior impacts the quality of your sleep life and optimize accordingly. Cross-referencing your data with that of others enables you to identify optimal behavior and consumption patterns that are not currently part of your routine. In this way, the system as a whole learns more quickly and efficiently because of the vast amount of data available from which to extract patterns.

Tools for optimizing weight and fitness include a scale like the one at withings.com that transmits your weight, body mass index and bone density to the Internet. You can tweet your weight and get support from the Twitter community to meet your goals.

Activity monitors that transmit activity data wirelessly to your iPod (like Nike's in-shoe unit), or fitness machines that transmit data to the Internet like those enabled with the iFit device, can archive all this data into a single interface along with calorie consumption and nutrition analysis.

Applications like Cal2Go aggregate all this information into a database where you can pull up visualizations of your consumption and activity patterns over time correlated to fluctuations in your weight, and relative make-up of fat, water, muscle and bone. Cal2Go cheerfully promises it "points honestly to weaknesses and progress" and gives you tips for meeting the goals you have set. If you want to see how much of your calorie budget you have left for the day, the application will tell you. You can invite up to five others to join in your Cal2Go community for support and accountability.

Fitbit is more of an all-in-one bio-monitoring system that tracks your activities throughout the day, syncs the data to your computer and—if you would like—shares it with your online community.

The Fitbit Tracker can sit in your pocket, or you can clip it to your clothes or strap it on your wrist when you are sleeping. Your number of calories burned, steps taken, distance traveled, and overall quality of sleep are sent to the Fitbit website where you can log in to see the compiled data and share that with your social network.

In the examples just mentioned, bio-monitoring interfaces still require relatively clunky devices or a lot of manual input. The next generation of bio-monitoring embeds sensors directly into the clothes you wear.

Imagine you are walking down a deserted alley in Rome after dark. Your jacket senses that your heart rate is increasing and your breathing is becoming slightly more rapid. You are beginning to sweat just a bit more. "Hey honey," you hear spoke softly behind you, "I am here waiting for you at the restaurant. Can't wait to see you!" You very well know nobody is really behind you, but the familiar voice is comforting just the same. At the same time, the jacket gets just a bit tighter and warmer, mimicking a hug. Ambient music begins to play and slowly fades as your heart rate and breathing return to normal. You have just had a mood altering experience—through your clothes.

Researchers at Concordia University and University of London's Goldsmiths have developed a working prototype that could give you just that kind of mood-altering experience. The clothing is embedded with bio-sensors that monitor your mood and give you feedback to alter it. "When you first wear the garment, you turn on the device and you tell it what person you want to channel that day," said Barbara Layne, professor at Concordia University and co-developer of the garments. "That could be your lover who's away; it could be your deceased parent, your best friend, whoever you want to be with that day." The multimedia is pre-loaded into a database for each person the wearer wants to virtually hang out with. "Multiple times during the day, you can set it for as many times as you want, and it will take your biometric readings, your bio-sensing data, analyze it on that emotional map and then go up to the Internet, to the database that relates that emotional state, and bring you back something that you need," Layne said.[6]

Monitoring and altering moods is only one use of bio-monitoring technology. Researchers at the University of California San Diego have created chemical sensors that can be printed directly on, say, the elastic waistband of underwear to monitor and respond to your state of health. Say you are a forest ranger monitoring fire danger and conducting research in a remote outpost in Yosemite National Park.

UCSD's bio-monitoring underwear video

One of the downsides of your choice of work is you are deathly allergic to a number of insect stings and bites. There is a good chance you would not have time to administer a shot of epinephrine quickly enough after a bee sting to avoid having your throat close and your airways cut off. Fortunately, the bio-monitoring sensors embedded in your waistband detect the change in your body chemistry almost before you can register the sting. A shot of epinephrine is automatically triggered, giving you time to scrape off the stinger and seek further medical attention.

By monitoring components of body fluids like sweat and tears, the sensors identify relative composition of bio-markers like lactate, oxygen, norepinephrine and glucose. The textile sensors can then diagnose injuries and identify changes in a patient's health status. Based on the diagnosis, the sensor system can administer the right drug automatically, stabilizing the patient while he awaits further medical treatment.[7] Bio-sensor technology will be valuable in arenas where medical intervention can be delayed, as in wartime, remote areas or places where there is a high risk of natural disaster.

Wearable sensor technology extends beyond clothing. Smart contact lenses that are embedded with sensors to monitor intra-ocular pressure and transmit status wirelessly to a receiver for the early detection of glaucoma are consumer-ready. Developers expect to receive approval from the United States Food and Drug Administration in late 2011.[8]

Other uses of smart contact lenses include those that use nanotechnology to monitor glucose levels in eye fluid and change color as an indicator alerting diabetics when blood sugar levels are getting low without them having to test their blood directly.[9] Such technologies are a little farther off, but early prototypes are currently functional.

The feasibility of always-on monitoring, transmitting and archiving of your bodily states will increase as the barriers to doing so disappear. Wearable monitors embedded in your clothing are one step closer to making the interface between you as a set of activities and the Internet seamless. Contact lenses are the next frontier—we are already accustomed to placing foreign objects on our eyeballs. As the technology for embedding contact lenses with tiny computers that can transmit information improves, the friction between always-on monitoring, and the interfaces to do so, will be dramatically reduced.

Genetic testing

My sister Renee and I have long known we are about as different as two people can be and still be siblings. We do share a love of learning new things, however, which made us both interested in using 23andMe's (23andMe.com) genetic testing service a few years ago. I was curious to learn how the process worked and my sister wanted to be proactive in her health care. We both took the test. When results came back a few weeks later we compared our genetic data. While there was little doubt before or after the test that we share the same set of parents, the differences in our respective profiles were illuminating. Not only did we have surprisingly few genetic traits in common according to this particular test, we were already experiencing some symptoms in our current health that 23andMe predicted might show up based on our genetic traits.

These kinds of tests are still very new. Standards for accuracy and protocols for testing are being scrutinized and refined. Still, the aggregate of genetic data undergoing analysis by companies like 23andMe is already yielding insights into drug sensitivities and disease risks. Comparing these data with bio-monitoring and behavioral data that will become available as more people participate in life monitoring, and new patterns of correlation between behavior and biology will emerge. These correlations can then be tested to determine which may be causally related and new treatments or behavior modifications can be devised.

Until now, research and analysis of genetic, biological, and behavioral data have been so limited they almost do not count in comparison to what is coming. In the absence of 24/7 monitoring, doctors learn to adjust patient self-reports of their exercise and consumption habits to accommodate for inaccuracies. Studies taking place in a lab are affected by the lab setting and therefore do not reflect real-world behaviors. Until recently, the time and resource costs to decode genetic information were prohibitive. That all changed in 2006, when public-private collaboration, increased computing power, and improvements in sequencing techniques sped up the process until the full human genome was published. That same year, 23andMe was founded and began to offer low-cost genetic testing and analysis to the public. Couple this with the pioneering activities of the self-quantifiers, connect it all to the network, and you get a game-changing landscape for discovery.

The amount of data entering the infosphere is only now beginning to ramp up. In a few years, this deep, tangible self-awareness will almost certainly have a revolutionary impact on what you believe to be humanly possible, including how long you can plan to live.

Nanobots and other inorganic agents

Nanobots and other inorganic agents transgress the biological-technological boundaries and may one day become active members feeding information into the social network.

Nanobots that kill cancer cells are already in proof-of-concept stage. The California Institute of Technology (CalTech) reported that researchers can now inject nanoparticles directly into a patient's bloodstream that are able to locate and attach to tumors and deliver "code" that can turn off a key cancer gene. The CalTech team demonstrated they can inject nanoparticles into the bloodstream at high enough doses to constitute a viable cancer therapy.[10]

Other inorganic agents include plastic antibiotics that work just like biological antibodies. Yu Hoshino and Kenneth Shea from the University of California, Irvine, in collaboration with researchers at the University of Shizuoka, Japan and Stanford University, made plastic nanoparticles that latch onto antigens just like natural antibodies. "These results establish for the first time that a simple, nonbiological synthetic nanoparticle with antibody-like affinity and selectivity (i.e. a plastic antibody) can effectively function in the bloodstream of living animals," they reported.[11]

While you may be used to the idea of medical mechanical intervention through surgery, chemical intervention through drugs, and biological intervention through vaccines, nanobots are a whole different animal—so to speak. Nanobots are inorganic autonomous agents smart enough to perform microscopic manipulations in the human body. At what point do these autonomous agents replace significant bodily functions or significantly augment or change genetic code in human cells? The answer to that question depends to a large extent on the kind of future *you* want.

Growing organs

The technology to grow or print organs from scratch is worth noting as it has far-reaching implications for your health and longevity.

One of my friends—I will call him Bob—is typical for his generation. Bob was born in 1951 in the San Francisco Bay Area into a fairly conservative household. By the time Bob was 16 or 17, he was ready to rebel and found he had lots of company. Teenagers converged on the San Francisco scene full of idealism and energy—and plenty of sex, drugs and rock and roll. Bob dropped out of high school and hit the party scene hard.

Eventually Bob grew up a bit, left the wilder side of life, married a nice British girl and started his own construction business. Like most of his peers, he continued to use drugs recreationally, and alcohol remained part of normal everyday life. Fast forward to 1998 and perhaps not surprisingly, Bob finds out he has cirrhosis of the liver.

The following decade was touch-and-go. We were never sure whether Bob was going to make it to see his next birthday. He was on a waitlist for a liver transplant, which is an iffy prospect. Getting a new liver depends on finding someone who is a match dying within reasonable proximity and being at the top of the list to get that liver. Bob had gotten his hopes up half a dozen times, only to be disappointed when some detail crossed his name off the list for that particular available liver.

Finally, good news—a liver that is a match becomes available in Florida. Bob finally gets his transplant and returns to full health. He now figures he has a good 30 years of liver function in him, so has gone back to drinking. "What the hell, life is short," he says, "so my liver will kick off when I'm 90—if I live that long, and by then there will probably be a new fix." Whatever you think about Bob's attitude, he is probably right about his prognosis for the future—assuming something else does not kill him sooner.

Maybe Bob only needs to survive another decade before he can order up new livers as needed. Researchers at the Center for Engineering in Medicine at Massachusetts General Hospital are optimistic that liver regeneration is within reach. Medical researcher Basak Uygun sees stem-cell science becoming advanced enough that you could donate your own cells which could be used to grow a new liver before long. Scientists would take your cells, differentiate them into liver cells, and create a scaffold by using a donor liver and carefully stripping it of its original cells. Your cells would then be spread over the donor liver scaffold, which is then used for growing your new liver. There would be no issue of organ rejection, because the new liver would be automatically compatible with your immune system. Uygun and team believe they should have working livers transplanted into rats within a couple of years. Assuming nothing goes wrong, he says, "we're hoping it will be in the clinic in five to 10 years,"[12] meaning that fully functional livers should be available then for human transplant.

Along with resuming his drinking habit, Bob has started smoking again, too. If his lungs become diseased, he may be in luck on that score as well. Using technology similar to the liver regeneration technique, in 2008 researchers from the University of Bristol and the University of Barcelona successfully replaced a main air passage in the lung of a 30-year-old woman. Scientists had grown the new air passage from a decellularized donor air passage scaffold of hard tissue seeded with the woman's own cells. The replacement airway ended up being fully functional.[13] Biomedical engineer Laura Niklason at Yale University and colleagues say they have used a similar method to produce a full set of rat lungs.

The team started with adult rat lungs that had all their cells removed, leaving just a scaffold of hard tissue. On the scaffold, they added lung cells from newborn rats. Niklason says "the crucial step was nurturing the would-be lungs in a bioreactor that circulates fluid—simulating what would happen during fetal development—or air through them. The cells stuck to the scaffold in the right locations and multiplied. After up to eight days in the bioreactor, they had coalesced into what the researchers' tests indicated was functional lung tissue." Due to the complexity of lung structure, replicating this process for successful transplant in human beings may be 25 years away.[14] Still, if Bob's body hangs in there, he may be able to upgrade to a new set of lungs before long, too.

How about if Bob is able to hang in there for even longer—what might be available in, say 30 or 40 years? Like print-on-demand books today, how about print-on-demand organs? Dr. Gabor Forgacs of the University of Missouri is working with a team of researchers to create bioprinting machines and technology that have already successfully printed blood vessels.[15] This 3D printing technique deposits human cells according to a design directed by a 3D computer model to create a living structure. The mechanics of printing more complex organs appear to be feasible, but so far printing hearts that work like the organically grown version may be decades away. See the video to get a glimpse of the 3D bioprinter in action.

Organ printing video

3D bioprinter video

You, but better

Sensors that transmit biological data to the Internet are

merely the beginning of the convergence of social media and technology. Boundaries between biology and technology are eroding as we speak. We are becoming transformed by technology as much as technology is being transformed by us. As the biological-technological feedback loops become more intricately connected—and as information passed between them speeds up and barriers to change disappear—the possibilities for what you and I consider "better" open wide. You now have the opportunity to begin imagining what you would like the quality of your life to be beyond the constraints of what you may currently perceive as physical limitations. Now that you know what is possible, how does that change your life?

CHAPTER 4

Your Daily Life: Smarter and More Efficient

Closing the gap between what you can envision and having it in your hands has the potential to rewire the brain in ways none of us can yet foresee.

Consider the fictional near-future scenario as described by our future protagonist, Jennifer:

I wave my hand to the right to flip through the Pandora® cover display in front of the windshield. Now that I am back on the grid I can finally chill out for the rest of the ride. This morning I put on the EmotionWear shirt and jeans I just bought—dying to find out how the latest version is. I flick my finger when I get to the TrueUtopia immersive entertainment experience to turn it on. Within a couple of seconds, my body blows away into a field of colors, sensations, and music as I head down the road toward my daughter's house. My eyes may be on the road, but the rest of me is definitely elsewhere.

A too-loud social announcement suddenly interrupts TrueUtopia: "Hey, Jennifer, only seven more for a swarm and we'll all get half price on any plant!" This is one of the things that make me ambivalent about always being tracked and connected to the info cloud. I could opt out of geo-social targeted announcements for "a small monthly membership fee" to Pandora, but sometimes they're useful—like now.

This morning I'd updated my status saying that I'd be visiting my daughter Isabella. The cloud knows Isabella likes to garden. It knows I am Isabella's mom. It knows I am six kilometers from Isabella's house. It knows Plantworld is just ahead. It knows Plantworld is offering half off all purchases if they can get 25 people to swarm within the next ten minutes. It knows eighteen people have already arrived at Plantworld. The cloud instantly crunches all this data and dispatches the geo-social targeted announcement I just heard through Pandora. I am amazed, a little annoyed, and also sucked in by the prospect of getting a plant for my daughter for half price—thus the ambivalence. "Oh well," I decide, "it's on the way. I'll just run in quickly."

I arrive at Plantworld and wave my index finger in front of a plant I like. An image pops up to augment the plant so I can see what it looks like in full flower: it's got a beautiful shade of crimson red. I pick it up and walk out the door to my car. As I pass the checkout area, a receipt display pops up in front of my eyes verifying I just bought an azalea and for half off the regular price. I barely remember when I had to stand in line at a checkout. The transaction is automatically deducted from my bank account coded into the NFID (near frequency identification) tag I wear.

When I get to Isabella's house, she places the azalea on the sideboard and we head to the kitchen/dining room for lunch. She punches a code in the air and the printer starts whirring softly. On her kitchen counter is a smallish transparent box with a few compartments. The box is connected to the cloud and uses a preloaded food-safe "goo" to print out dishes, glasses, and utensils as needed. The code Isabella enters tells the printer which serving ware to select from her favorite designs and how many of what she would like printed. All the printing material is recyclable.

She starts to punch in a second set of codes, then hesitates—"I forget, Mom, do you like Japanese?" I tell her I love Japanese food—it's Chinese that I'm not so crazy about, at least as it tastes coming out of the printer.

What started as a molecular gastronomy fad a decade or two ago with breaking foods down into their basic components has evolved to the point of convincingly replicating the essential flavors and in-the-mouth feel of a cornucopia of cuisines—plus adding a bunch of taste innovations that require new words to describe.

While chefs were innovating food preparation techniques, scientists were developing food printers to automate home meal production. Not too long ago, the trends converged so that we now have the option of instant meals—most of which are really quite good. I still like to use raw ingredients on occasion, but cooking from scratch is more of a craft than something I would want to do for every meal. Now that printed food tastes great and you can get its nutrients personalized and optimized for your health, there is no real health reason to cook from scratch any more than you would want to make all your clothes from scratch, at least in my opinion. Some of my friends argue that the intentional preparation of food adds "love" to the final product, and maybe they are right. Compared with the so-called "fast food" of before, however, today's convenience food is apples to oranges when you taste its freshness and consider the relative health benefits.

After we finish lunch Isabella puts the dishes in a chute embedded in the countertop near the printer. The remaining food is separated from the dish material so it can be shot outside into the community compost bin. The dish material is disintegrated, sterilized, and prepared for reprinting.

As it turns out, washing dishes takes a lot more energy and resources than simply printing and recycling them as needed. Printable items include just about anything that can be designed. Designs for everything from prescription eyeglasses to toilet brushes are easily accessible from the cloud and can be customized before printing. Advertising meant to generate desire for utilitarian items is almost obsolete—why would it make sense to buy permanent things that can instead be made-to-order in whatever form you want—and instantly on-demand? Why take up all the space storing things like dishes and glasses as well as keeping them organized and clean, when you can make them as needed and recycle as soon as you are done? A few physical items are still worth marketers' time to advertise—like the azaleas on sale. Living plants are still easier to get in a store, and original art remains valuable, but other than that most everyday things are easy to print—even clothing, though not yet the most comfortable.

Isabella takes me outside to see the studio she is building in the backyard. Furniture is another item easier to buy from a store—unless you have a giant printer, which everyday people do not. Her studio is cozy even without a desk and chair—space is definitely at a premium. We bring up the IKEA store on my iPad and select a desk and chair to see if it will fit. Holding the iPad in front of where the furniture will go, we can see exactly what it will look like in the room.

IKEA's store application overlays a real-size 3D image on the physical environment making it easy to see how something will look without having to buy first. We try out a few configurations, find the one that looks best, and hit a button to order and arrange for pickup.

Virtualization and just-in-time production

All elements described in the near-future fiction story of Isabella and Jennifer are either currently available or in working prototype. Gestural user interfaces, sensation altering media, geo-social targeted advertising, augmented reality product enhancements and displays, chip-enabled instant payments, and the 3D printing of dishes, utensils, and food are all feasible today, at least in rudimentary form.

Starting off the list, one of the more well-known examples of a gestural user interface is the SixthSense project that generated *oohs* and *ahs* a couple of years ago following its introduction at a TED conference. Described as *Minority Report* brought to real life, researcher and developer Pranav Mistry demonstrated moving digital images in front of him by flicking his finger and taking a picture by simply creating a box that frames the shot with his thumbs and forefingers.

SixthSense demo at TED video

Robert Wang, a graduate student at MIT where Mistry does his research, points out that SixthSense and similar gestural UI prototypes currently only work in two-dimensional space. Describing low-cost gestural interfaces that use markers attached to fingertips to manipulate digital data, he says: "That's 2D information. You're only getting the fingertips; you don't even know which fingertip [the marker] is corresponding to." Wang and Jovan Popović, an

Associate Professor of electrical engineering and computer science at MIT's Computer Science and Artificial Intelligence

Laboratory, developed a new system that uses a multicolored Lycra glove to manipulate virtual objects in three dimensions. Wang explains: "This actually gets the 3D configuration of your hand and your fingers; we get how your fingers are flexing."[16] It is not a big leap from gestural user interface prototypes such as these to envisioning Jennifer flipping through virtual entertainment selections in her car or deciding to purchase an azalea appearing in full bloom in the store, or helping choose her daughter's furniture by way of an IKEA augmented reality application.

3D gestural user interface video

An even more efficient system for manipulating virtual information is one that requires you only to *think* about

what you want to have happen for it to happen. San Francisco-based Emotiv Systems sells a US$299 headset that lets you do just that. The headset has a dozen or so sensors that pick up the electrical activity your brain gives off as you think about performing an action such as selecting a document file to open on your screen or typing a computer command. It then interprets those electrical impulses into the actions you are thinking about performing. Researchers at Dartmouth have connected the Emotiv headset to a device that lets you control your smartphone— just think about calling your mom and the phone starts dialing.[17]

Emotiv's headset in action video

Thought-controlled smartphone video

University of Wisconsin researchers have

developed a brain-computer cap that allows you to send messages to Twitter simply by thinking of letters to spell out the words. It's not as impressive as Emotiv's device that recognizes whole words and can be trained to translate such words into action but worth noting because it may be applicable where simple letter selection is all that is needed. Manipulating virtual data through gesture or thought, however, falls short of creating an experience that convincingly replicates interaction with the material world. Gestural or brainwave user interfaces are

Twitter by thinking
video

perfectly fine for visual or aural navigation, but they do not engage the tactile senses. A three-dimensional visual display—a hologram—might be good enough to replicate the look of a physical object, but if you tried to interact with it, the illusion of it being real would quickly be broken. Hiroyuki Shinoda, Associate Professor at the University of Tokyo, has developed a technology that may bridge this gap between illusion and reality. "Up until now, holography has been for the eyes only, and if you'd try to touch it, your hand would go right through," he explained to Reuters. "But now we have a technology that also adds the sensation of touch to holograms." The technology uses ultrasonic waves that create a sense of pressure when you reach out to touch the hologram.[18] In addition to holodeck-like entertainment applications, replacing physical objects with virtual ones could have practical value. Virtual objects may be produced at lower cost and they can be used indefinitely without wearing out. In settings like hospitals, minimizing contact with surfaces like a light switch or water tap may help maintain sanitary conditions as well.

While holographic virtual objects have their practical applications, they cannot fill every role. Sometimes you need an actual physical object—like food to consume, plates to put it on and utensils to eat it with. 3D printers are rising to fill this need. MIT's Cornucopia digital gastronomy project has developed two working prototypes for the 3D printing of meals. The Digital Fabricator is intended for home use. It stores and mixes ingredients, which it then cooks or cools as it "prints" the mixture onto a surface. The technique allows complex combinations of ingredients to be prepared and deposited precisely by an extruder. According to project leaders, this process "not only allows for the creation of flavors and textures that would be completely unimaginable through other cooking techniques, but, through a touchscreen interface and web connectivity, also allows users to have ultimate control over the origin, quality, nutritional value and taste of every meal."[19] Because it requires thoughtful selection of ingredients, the Digital Fabricator is perhaps more of a precision cooking gadget for cutting-edge chefs than a way of making 3D printed nutritious and delicious home meals fast and easy—for now, at least. As a technology geek who loves to cook, I for one am interested to see what currently unimaginable flavors and textures the Digital Fabricator comes up with.

Cornell University is attempting to resolve this problem of ease through what its developers call a "Solid Freeform Fabrication" (SFF) of food. Unlike MIT's Digital Fabricator, which depends to a large extent on culinary know-how, Cornell's SFF aims to use a combination of xanthan gum and gelatin as a base, along with flavor agents, to simulate a broad range of foods as easily and with as few ingredients as possible. They envision that the know-how of the best chefs and information from leading nutritionists can be "abstracted to a 3D fabrication file and then used by laypeople to reproduce famous chefs' work in the home [and] to help laypeople eat more healthily, without necessarily having to learn healthy cooking techniques or even understand nutritional principles such as caloric intake and protein balance."[20] The SFF files could be customized for your personal nutritional needs. As described in the previous chapter, "Your Massively Networked Self," the SFF could be tied into all the data you are providing about your workout routine and health stats. When you arrive home from the gym, a tasty meal could be waiting for you, specifically tailored to your calorie and nutrition needs based on your workout.

Cornell's SFF is still in early stage development. With advances in 3D printers and experiments in molecular gastronomy progressing, the promise of easily printing delicious, nutritious meals in the home is quickly becoming feasible.

Printing plates and utensils from computer-generated designs is already here. RepRap (short for "replicating rapid prototyper") and MakerBot are perhaps two of the best-known 3D printers in existence today. Each uses open source technology and works by extruding a material onto a surface, adding tiny layers to copy the form of a computer generated model. Practically anything that can be modeled can be printed. The current limitations are only the size of the printer itself and the suitability of the medium each printer uses for what you want to create. RepRap uses polycaprolactone and MakerBot uses ABS plastic as the printing medium, so assuming you are okay with plastic plates and utensils, you can go ahead and print your tableware right now—with an investment of a little under US$1,000. CandyFab is another 3D printer in early stages of development that uses granulated sugar instead of plastic. According to project leaders, sugar is a low-cost production medium that makes rapid prototyping even less expensive than with plastics. Fab@Home is yet another open-source 3D printer that is so far unique in that it can extrude a variety of materials—from silicone and cement to cake frosting and cheese.

Virtualization of shopping has in some ways been around since the Sears, Roebuck and Co. catalog came on the commercial scene in 1888. In the late 20th century, online commerce made the virtual shopping experience even more robust—and now it is getting better with applications that use augmented reality technology. Swedish home furnishings maker IKEA launched a limited release iPhone augmented reality application that allowed potential buyers of its novel PS line to see how pieces would look in their homes. Once buyers downloaded the application, they could point the iPhone to a section of the room they were looking to furnish and select a piece of furniture. The furniture image would automatically scale to the room so the buyer would have an accurate idea of placement. If you wanted to share with a friend, the application allowed you to take a snapshot of the scene.[21]

Ray-Ban's Virtual Mirror augmented reality application allows you to try on different styles of glasses right from home using a webcam.[22] The U.S. Postal Service has a Virtual Box Simulator augmented reality display so you can select the correct size to mail your contents.[23] The possible applications for further virtualizing shopping are limited only by retailers' imaginations.

Geo-targeted just-in-time advertising is also feasible now, made possible by combining the technologies of location-based services like Facebook, Foursquare, and Gowalla with group-based coupons such as those provided by Groupon and geo-targeted advertising. Facebook keeps you signed in and tracks your movements around the web currently, unless you opt out of that feature. If you have Facebook on your smartphone and have the GPS turned on, Facebook can track your physical movements in space, too. Facebook partners such as the Pandora music site—one of the companies to participate in Facebook's "Like" sharing platform pilot project—potentially have access to this data as well. Facebook's platform-enabled social graph combined with the personal information you choose to share and the ability to track your movements online and off, coupled with the potential for sharing this information with an advertising-supported service like Pandora, and you are now fully capable of receiving the instant announcement that you can get half off at a Plantworld around the corner, *if* you get there right now along with 25 others. Welcome to just-in-time marketing.

Regardless of the final form any of these take or if those in prototype become commercially available, there is good reason to believe that the trend toward virtualization and just-in-time production will continue to ramp up. Convenience is one motivation; retailers' profitability is another, and the technology will only become more sophisticated. Even more optimistic are the potential ecological benefits arising from 'on demand' production, as we will now examine.

The end of mass production

Commercial mass production of goods is ecologically inefficient on many fronts. It requires that goods be warehoused until they are sold to a retailer. Once sold, they must be transported to a store for retail sale. There they are stored once again by the retailer until they are sold to the individual consumer. The individual consumer must then transport these to a home where they are stored yet again in between uses. Mass producing, repeatedly storing, and transporting all these goods represents an enormous waste of space, time, and energy. Moreover, once the product has outlived its usefulness to the consumer, it is placed in the trash or recycling bin. Once again it is transported to either be stored in a landfill or, for a very small percentage of waste, to be recycled. The finite resources that went into making the product are, in the end, mostly wasted.

Virtualization, just-in-time production, and instant recycling eliminate waste and conserve resources. How much smaller could your kitchen be if food were virtualized into a tasty, visually attractive, nutritious hydrocolloid form? "Hydrocolloid," in which microscopic particles are evenly dispersed in a water base, may not sound that great, but I expect marketing would take on the challenge of creating a positive brand association with the product. You wouldn't need to store dishes, utensils, raw ingredients for meals, frozen fast foods, canned goods, or pots and pans. Would you still need a large stove and oven? Probably not. What use would a refrigerator large enough to crawl inside and close the door be? You would doubtfully need all that cold storage space, and you would surely think of many other uses for the freed-up space in your home.

Similarly, if you could print clothes on demand in great fabrics and in the latest style, designed to fit your body perfectly, would there be any need for a huge walk-in closet to store a bunch of clothes? Freedom of Creation in Amsterdam and Philip Delamore at the London College of Fashion take your three-dimensional body data and create 3D printed pieces of clothing. As with 3D printed food, clothing can be tailored specifically for the wearer. Moreover, the materials can be melted down and reformed into a new outfit.

The same argument could be made for many utilitarian items taking up space in your home. Not only that—what do you do with all those items once they wear out or become obsolete? Meal production from scratch may go the way of artisan craft, with stoves, ovens, and cooking gadgets becoming tools for hobbyists just as sewing machines are today.

As technologies for 3D printing and recycling improve, transportation costs approach zero, material waste is reduced, and the need for storage may begin to vanish. The feedback loop between what you want and getting it in your hands is dramatically shortened. The benefit of making the loop shorter is that production becomes more efficient and less wasteful. Much as the advent of the Internet closed the *information* loop connecting what you want to know and knowing it, this next wave of *material* innovation will close the loop connecting what you want physically and having it in your hands.

Grow-your-own groceries

While print-on-demand prepared food could fill some needs, there is no good substitute for fresh-picked lettuce when you want to have a salad. Meeting the goal of growing groceries while keeping the costs of production, time, and labor down are how we got from small village or family farms to big agribusiness. Big agribusiness developed technologies to make feeding large numbers of people efficient and cost-effective—with some unintended long-term consequences that we now see as ecologically unfriendly. For instance, farm technology became more efficient at controlling the local environment to maximize productivity by creating uniform crops and improving pesticide- and weed-reduction techniques. The introduction of "Roundup ready" crops that are genetically modified to tolerate the herbicide Roundup, for example, meant the cost of keeping crops weed-free dropped—for a while. That is, until Roundup-resistant super weeds that can choke crops grew onto the scene.

To maintain the health of the desired crops, new technology must be employed to keep control over the environment or the crops will fail. In the case of super weeds, farmers had to go back to using a variety of pesticides and incur greater costs. Big agribusiness is now so dependent on technology to control the environment that maintaining the technology gets all the focus and the initial goal of feeding people somehow gets lost. Some farmers have taken a step back and realized the current system wasn't working, so they created organic farms that include a diversity of interdependent crops and work with the natural environment to produce crops. This is perhaps a step in the right direction—though a less cost-effective and more labor intensive one.

There is no single right answer to the question of how to feed people. The best answer emerges from the needs of the community needing to be fed and the environment of which that community is a part. If we were living in a village in the 18th century, having members of the community who specialize in producing vegetables or dairy products or meat would make sense. The ecosystem is small enough and its members connected enough that information in the feedback loop is able to remain well connected. The return to artisan production in the 21ˢᵗ century, however, is not an optimal solution—though for those who choose to spend their time doing so, it can be a viable option. Those who advocate for artisan or localized production are on the right track from the standpoint of creating a more intricately connected ecosystem. However, there is a reason big agribusiness became the standard for food production. It requires infinitely less time from individuals in the system as a whole and everyone still gets something to eat. Community gardens are time-consuming to run, and the success of these small farms rely on having enough people interested in making farming their livelihood. Furthermore, even when small farms have enough people interested in tending them, they have a tough time remaining profitable enough to survive. The ideal situation is to produce food locally with minimal human effort and at low cost. Until recently, that seemed like a request that would be hard to fill.

One option may be to create small, automated remote-controlled farms that leverage the latest in robotic technology. MIT's Computer Science and Artificial Intelligence Laboratory (CSAIL) has developed a working mini-prototype of an automated greenhouse tended by robots and filled with communicating fruits and vegetables. These plants are equipped with networked sensors that calculate soil moisture and nutrients, track their fruit production, and communicate these data to the network. The robots are also connected to the network so they can pick up communication from the plants and respond accordingly. They have arms equipped to deliver water and food to plants that need them.

Robotic farmers video

They are able to find a specific piece of fruit to harvest, they can pollinate plants, and they can fulfill other plant-tending duties like identifying those that need to be weeded out.[24]

The fact that the CSAIL ecosystem is regulated by a complex set of instructions that regulate its health and automate its function is impressive from a computing perspective, if not unprecedented. Factory automation to produce everything from cars and candy has been around in some form for many decades, all the while becoming increasingly sophisticated. What is unprecedented about projects like CSAIL is that the communicants in this case are living plants and intelligent machines. They are speaking to each other within a largely autonomous ecosystem. Once put in motion, they are part of an ecosystem that may be able function and produce new life without much intervention from human agents—at least for as long as its components function as planned. You can imagine, perhaps, small community farms dotting the landscape run by robots while community members take turns monitoring and tending via a remote-controlled iPad application.

Another option is to create mini ecosystems that rely on their inhabitants to tend each other. Sweet Water Organics in Milwaukee is a bootstrap urban fish and vegetable sustainable factory farm. In 2008, they took over an abandoned industrial building and turned it into a proof-of-concept farm to demonstrate creative recycling of the urban landscape and a profitable contribution to the local economy. Tilapia and perch waste fertilizes the varieties of lettuce, herbs, tomatoes, peppers, chard, and spinach the farm grows, and the vegetables filter the water for the fish. The founders want to reclaim unused urban "deserts" to provide healthy food and create jobs for local residents.[25] Both the CSAIL and Sweet Water Organics projects take a step toward creating self-sustaining, organic ecosystems that meet the need of localized food production while making that process less dependent on manual labor.

The efficiency advantage

If you could reliably get what you need just when you need it, how do you use the freed time and mental energy that was formerly dedicated to gathering resources to acquire food, clothing, tools, and the like? This is already happening to some extent as folks downsize material possessions and rely on sharing sites like ToolSharing.com for access to stuff they need when they need it so they have more space, more time to travel or do other things they enjoy. Do you, like downsizers today, take advantage of that freedom to participate more in your community or see places around the world you have never been to before? Or, do you become so fascinated by the explosion of creative possibilities that you begin to suffer from the analog to information overload—that is, creation overload? I suspect you will see a range of responses to any one of those questions depending on the individual.

Virtualization and instant personalization of your shopping experience has its potential efficiencies as well. Like the life-size IKEA couch you can see in your living room before you buy, or the styles of Ray Bans you can try on virtually before placing an order, shopping now comes to you—you don't have to drive, bus, or walk anywhere. "Going shopping" becomes an anachronism. The experience of purchasing something only to find when you get it home that it doesn't fit or look quite right becomes a thing of the past.

Virtualization also has advantages for marketers, for better or for worse, depending on your point of view. All that virtual data you feed into Facebook, Twitter, Foursquare, Gowalla, LinkedIn, and 23andMe among other social media applications yet to make the scene is already being connected within a map of relationships, entertainment preferences, behavioral patterns, location, political views, sexual orientation, you name it—the list is exhaustive. Couple that data with increasingly smarter algorithms, and marketing analysts can predict how likely you are to respond positively to a particular offer. Add yet another layer of gaming dynamics that use what marketing analysts know about your current patterns of behavior to persuade you to adopt new behaviors, and you begin to see why becoming conscious of what kind of world you would like to create is increasingly important.

Jennifer, Isabella's mom in the story, views the Plantworld group coupon as simply a convenient benefit of contributing to the information cloud. Would she have purchased the azalea had she not been sucked in to the game of getting to a destination within a few minutes for the chance to win a prize of 50% off? Groupon works in exactly this way to great effect. Creating time-sensitive and location-dependent conditions for winning a prize are classic elements of a game. When The Gap created a Groupon voucher for consumers to get US$50 worth of Gap merchandise for US$25, sales hit US$11 million in a single day.[26] Groupon received a significant portion of the take—estimates range from US$4 million up—and that commercial success only used the game dynamic. As companies successfully merge sophisticated behavioral data with persuasive gaming techniques like these, the promise for sales is almost guaranteed.

Persuasive technologies like these are neither good nor

bad in themselves—it is what they are used for that makes them potentially positive, negative, or neutral. That marketers are becoming more sophisticated in their use is perhaps no surprise. What happens when the very home you live in becomes part of the persuasive fabric of your life?

The new smart home

The smart home of the past usually referred to home automation: lights automatically going on and off as you enter and leave a room, ambient music adjusting to the preference of each person, refrigerators sensing when you are out of butter, and so on. These examples all fall in the realm of 'smart' only in the sense of being programmed to adjust when a specific set of conditions arise. This kind of smartness is little more than predetermined input-output. The next wave of smart home is something else altogether—it can learn and actually interact based on the information it is absorbing and making sense of.

Washington State University's CASAS Project is developing intelligent environments that you can truly consider smart.[27] Prototype homes are embedded with sensors that track behavioral patterns of both the environment and its residents. Information from these sensors is fed into a processor that maps the events according to what triggered them and uses a set of meta-instructions to make sense of these patterns of behavior. The data is collected from multiple residents, and each resident is recognized as an individual. The meta-instructions, or algorithms, create a responsive feedback loop that observes how individual residents tend to behave, and monitors those behaviors to continually adjust and learn from their patterns.

Over time, the system is able to create a living environment that is responsive to the needs of individual residents.[28] Residing in an environment that can track your behavior, map which series of behaviors go together in a typical sequence—like getting up, starting coffee, getting dressed, and eating breakfast—and learn what those behaviors imply is a first step toward eliminating the lag between you wanting a cup of coffee and that coffee appearing. Working prototypes of intelligent homes like these exist today.

The Happylife project is an experiment in reading the moods of home inhabitants. It aims to use real-time data from facial expressions, eye movements, changes in pupil dilation, and body temperature, among other data points to ascertain moods of individual family members. Each individual is asked to respond to a series of scenarios in order to set a baseline for a variety of emotional states. This data is coupled with computer models that map facial expressions to psychological states along with other existing behavioral analysis models. Jimmy Loizeau and James Auger built the visual display linked to a thermal image camera that uses facial recognition to differentiate between members of the family. They explain: "Each member has one rotary dial and one RGB LED display effectively acting like emotional barometers. These show current state and predicted state, the predicted state being based on years of accumulated statistical data."[29]

Applications of smart home technology are not limited to anticipating that your kid is about to take a shower and heating the bathroom to the temperature she or he prefers, or to giving you feedback on changes in your current psychological state. It extends to objects in the environment. Adam Lassy (www.adamlassy.com), a software engineer and designer, transformed standard pieces of IKEA furniture into responsive agents. Lassy envisions producing autonomous furniture that adapts to your needs. When you walk into a room, furniture moves out of the way and moves back again after you pass. When you are having a party, the furniture circulates to deliver drinks and snacks without interfering with the flow of traffic Extend those ideas out a few years, and formerly inanimate objects may appear to anticipate your needs and respond proactively. The technology already exists. The potential for you to set up the conditions for your home environment to learn about specific behaviors you may not be aware of yourself, and identifying ways to modify those moods or behaviors so you can achieve a particular goal may not be far off. Who knows? Maybe one day your home notices that you are feeling less happy than usual for a few days. Does it trigger a message on your bathroom mirror suggesting you would feel better if you went for a run or called your friend Ben?

Responsive furniture video

At a deeper level, living in a home that interacts with you in tangible ways may be the first step toward raising consciousness about how everything around us already influences our behavior. I divide my time between living the city life in San Francisco and living the country life in rural Montana. I choose to be in one place or the other depending on the environment of influence that is best suited for what I want to accomplish. Montana is a great environment for literally inspiring blue-sky thinking. It is easy to gain broad perspective as I look out on views of endless blue sky obstructed only by distant mountain ranges. San Francisco is a great environment for creating structure and practical planning. Surrounded by buildings, navigating a structured network of streets and interacting with dozens of people every day requires me to continually define my actions within changing contexts. In the city, I need to account for the Giants playing at the ballpark, which means lots of traffic to and from South Beach. The best route from among the many I could take to get to the dog park to walk my Tibetan Terrier, Tika, is one that circumvents this traffic. My days in the city require constant decision-making based on a large variety of changing conditions. The influence of the city is that it continually keeps my mind in a state of practical planning—which is great when I need to be in that mode, and not so great when I need broader perspective.

If you're like most people, you currently take your environment for granted and remain largely unaware of its influence on your life. As living spaces become smarter and more interactive, you can become more aware of the informational environment you reside in and adjust it more precisely to the conditions you would like to create. The feedback loop connecting your environment to your behavior is probably not very tangible and measurable at this point. The more tangible the connection to your living environment becomes, the more aware you will be of how parts of your environment affect you and, as a result, the more you will be able to change aspects of it to make you happier, healthier, and more productive.

Questions to consider

When you can near-instantly manifest everyday goods, do they recede into the background as something you no longer have to think much about? Does the materialistic quest to hoard and acquire lots of stuff go away? When you can reliably get what you need just when you need it, how do you use the freed time and mental energy that was formerly dedicated to gathering resources to acquire food, clothing, tools, and the like? When you live in an environment that actively responds to you, does your perception of your place in the universe change?

CHAPTER 5

Rethinking Your Worklife

Getting a college degree, embarking on a career, saving money for retirement and retiring at 65 used to be considered a sound strategy for living a good life. Recently the assumptions underlying that strategy have been called into question. Fortunately, you now have the tools to help you start conceiving alternatives.

When are you most happy during the day? If you live in the U.S. and said "in the afternoon" you are likely in the minority. Northeastern and Harvard University mined U.S.-based Twitter posts for three years of mood related keywords broken down by hour. Their study found the most disgruntled, least cheerful posts consistently concentrated between noon and 4 p.m. each day. When are most U.S. residents in a good mood? The best moods pop up in the morning before 9 o'clock and in the evening after 8 o'clock.[30]

Since most employed people work daytime hours, it is reasonable to suspect you are happiest before you go to work and after you leave. Why would so many of us choose to spend so much of our time doing something that—at least according to this analysis of social media posts—makes us unhappy? Do you have a choice?

Twittermood graphic

The happiness factor

Madison graduated from UC Berkeley with both bachelor's and master's degrees in Architecture. She worked hard throughout high school to be one of the lucky few selected for admission. Once a UCB undergraduate, Madison stayed focused enough to build a great academic record that got her into a very competitive architecture graduate program.

Driven by her dream of designing great buildings, Madison thrives in her studies. Like many aspiring architects, she wants to create fantastic structures that will bring the surrounding landscape to life. Frank O. Gehry's Hotel Marque de Riscal in Spain—with its undulating, multicolored surfaces that appear to have gently floated onto the landscape and seem they might float away again given a stiff enough breeze—is an inspiration. However, the reality of being a newly degreed architect begins to dawn on Madison as soon as she enters the workforce.

World-renowned architectural firm, let's call them Hahmer, Dillon & Mills (HDM), hires her to be a Junior Associate Architect. At first she is thrilled to have the opportunity to learn from such experienced and accomplished talent. She soon realizes that her expectations are not matching reality. She spends her days at the drafting board carefully putting the details of other people's ideas on vellum paper and has near zero interaction with the senior staff. She quickly understands she will not be sitting in the creative driver's seat any time soon—if ever. Assuming she chooses to continue working for HDM, she will remain a technically skilled cog in the building-making machine for years. She fears that by the time she has the chance to exercise her creative muscle, it will long have atrophied. So she quits.

Madison's parents are baffled by their daughter's decision. After investing more than a decade of effort and over $100,000 in her education they do not understand why she would want to close the door to the path that has been laid out for her. Her mom tries to convince her that the hard part of building foundational credentials and getting a foot in the door has already been accomplished: "Now you only have to prove yourself and you'll make great money along the way," her mom says. "At least stick it out for a few years," her dad says, "until you build a decent resume. Then you'll have the option of working for a smaller firm where you can have more creative control and still make money."

Something clicks. Madison realizes on a gut level that she does not ever want to work for the money. Her parents will no doubt think she is idealistic or irresponsible if she says so out loud, so she keeps that thought to herself.

It turns out that Madison is smarter than she may know. Studies demonstrate that once Madison has enough money to cover her need for sustenance and some freedom to enjoy her life, increasing the amount of money she has will bring less incremental satisfaction.[31] In other words, money does buy happiness up to the point where basic needs are covered. After basic needs are covered, increasing the amount of money you take in begins to yield less and less satisfaction. Not only does having a lot of money fail to make you happier, but other studies indicate that having a lot of money may actually make you unhappy. Researchers at the University of Liege in Belgium conducted a study to evaluate the correlation among joy, awe, and contentment (i.e., "happiness") and money.

They showed study participants scenes meant to inspire happiness, along with stacks of Euros—some small and some large.[32] Participants who saw smaller stacks of Euros while being asked to respond to a happy scene had higher levels of positive response than those who saw larger stacks of Euros. Not only that, but those who made more money in real life relative to the others scored even lower still on the happiness response index. The researchers hypothesize that participants who know they can have the best at any time are ultimately robbed of the ability to enjoy everyday, ordinary experiences.

Madison instinctively knows that her parents are wrong to assume making a good living is tied to earning a bigger paycheck. She is realistic about the need to earn enough to pay her bills and take care of shelter and food, and is confident she can cover those expenses well enough. More importantly though, she is fairly certain her parents are the last generation still able to take long-term employment with a company—or even in a single career—almost for granted. The marketplace is being disassembled and is behaving in increasingly unpredictable ways. Turning time into money through hard work is no longer guaranteed. Why would she keep doing something that makes her unhappy when there is little guarantee that her investment will pay off in happiness down the road? Madison knows something her parents do not. She knows she lives in a time when she really does have a choice.

When did life get so complex, anyway?

Madison is on the cutting edge of rethinking worklife. Since she is just starting out, she can choose to chart a different path for herself relatively easily. The following story about a modern couple, on the other hand, is probably more like the rest of us. James and Allison are enmeshed in the current economic realities. Making a change will first require that they become conscious of the assumptions that have been driving their lives to an almost unsustainable complexity.

James, Allison, and their toddler son, Malcolm, would have been well set for a comfortable upper middle class lifestyle a generation ago. When I first met Allison nearly twenty years ago, she was a couple of years out of college—having graduated debt-free thanks to an athletic scholarship for volleyball. She was working full time as a corporate sales associate, getting her pilot's license, completing an MBA degree and playing volleyball competitively all at the same time—a typical Type A overachiever. Allison's financial management skills were enviable. By the age of 23 she had a solid retirement portfolio and contributed a significant portion of her income to it each month. In the years that followed, she continued to excel in corporate America, building her skills, network, and resume. She is smart, highly motivated, a natural networker, and comfortable with promoting herself.

Nearly two decades later, however, she finds herself unable to make ends meet. She feels very fortunate that she started saving and investing for retirement at an early age, because her retirement savings has been the only thing keeping her and her family off the street.

Allison works in corporate executive travel sales where a single commission can bring in six figures. The base salary, however, does not amount to much. After the economic downturn a few years ago, corporate travel budgets were among the first to go. She found herself competing for sales in a price-driven marketplace with a travel product that had always touted its value based not on price, but on brand name recognition. Not surprisingly, sales trickled to nothing and she was left trying to live on a base salary that did not cover monthly expenses. In addition to her home mortgage and basic living expenses, she holds a mortgage for a condo she purchased for rental income when the Denver housing market was booming. The condo is now worth less than the total mortgage amount, and she has not been able to rent it recently. For the last couple of years, credit cards have been filling the gap between income and outgo while Allison diligently looks for other work. She recently reached a point where she realized she could no longer continue racking up debt to live. Then she got laid off from her corporate job.

Alison's husband James runs a home remodeling business. He met Allison at a friend's wedding and liked her immediately. Allison took some convincing at first, but he won her over once she got to know him. Even after the housing market downturn, James continues to work nonstop. He does whatever it takes to win the few available contracts out there—which often means underbidding the competition and cutting his profits down to almost nothing after he pays for materials and subcontractors. He has worked hard all his life and feels like he would lose a big part of his identity if he stopped trying to make his business successful. He is perpetually optimistic that it is about to turn a corner to profitability. In better times, his optimism would likely be warranted. For the last several years, however, profits have been almost nonexistent. James does not like to talk about it. He equates working hard with being a good provider, regardless of the economic realities.

James and Allison are two highly capable people willing to work hard and work smart. They brainstorm ideas for new business. Allison is helping James on the business side to make his business more profitable by extending its services to regular home maintenance contracts. James is helping Allison with ideas on branching out into consulting work. I know them fairly well; there is no question that they are doing everything humanly possible to make things work. In hindsight, they know they should probably not have relied on credit cards to fill the income/expense gap for so long. Had the economy turned around sooner, however, they probably would have been able to pay off the cards and not be faced with such insurmountable debt. The credit spigot is now closed, so they have no choice but to face the reality of their situation. Cards and credit lines are either maxed out, or credit card companies have reduced their available credit to $0.

They meet with a financial counselor to help them get a true picture of their financial situation. The counselor says they have no choice but to file for bankruptcy. Allison in particular has a hard time accepting the fact that she will not be able to pay what she owes. For her, bankruptcy is a moral issue, albeit less of one than letting the family starve. They file for bankruptcy, which fortunately clears most consumer debt. They still have food, medical, and car insurance, mortgage on their home, and car payments to make. Allison's $150K+ retirement savings are being drawn down prematurely and at a high penalty to keep the basics going.

Massively Networked

James invested whatever money he did have in the remodeling business long ago; he has no reserves. Although nobody would blame them for feeling depressed, Allison and James continue to maintain a solution-oriented mindset. They focus on the positives—they love each other, they have a beautiful son together, and they are both capable and healthy.

Stories of people like James and Allison are increasingly common today and might be similar to your own. If you are like Allison, you grew up assuming life would go along pretty much as it always has—work hard, get an education, save for retirement and invest in a home, and you would almost be guaranteed a roof over your head, food to eat, and clothes to wear. Today's reality is that many smart, hardworking people who did all the logical things are having a difficult time keeping their homes and feeding their families. James and Allison are still relatively young and healthy. For those who are nearing retirement the situation is more challenging.

One of my clients, a large brokerage firm, brought me on board to help develop an online community plan addressing the needs of those nearing retirement. The timing was critical because most had recently seen their portfolios cut in half in the economic downturn of fall 2008. We set up interviews with investors aged fifty and over to hear their perspective. Sam is only one of the interviewees, but his story is a common one among them.

Sam is a 62-year-old semi-retired IT consultant who purchased a home eighteen years ago in an up-and-coming East Bay neighborhood near San Francisco. A large part of his retirement fund was to come from the sale of his home, which he had planned to sell within the next year or so and downsize to a condo. He expected to pay for the day-to-

82

day expenses during his retirement from funds he would draw down from his IRA. As with so many others, his house value has dropped 35% in the last year, which amounted to a significant portion of its appreciation over the years. With the glut of housing on the market, he has had no reasonable offers. His investment portfolio had been cut approximately in half, like most of the others. Nobody was predicting a turnaround in either the housing or stock market any time soon—certainly not in time for Sam to fully retire at 65 as he had expected.

Sam, like many of his peers we interviewed, simply wanted to know what their near-term options were for keeping as much money as possible in his portfolio. He was also interested in learning how others are planning to manage the shift in expectations—pushing out retirement, cutting back lifestyle choices like eating out or traveling abroad, investing in health to avoid healthcare expenses down the road and even taking on boarders for houses it no longer makes sense to sell. The online community site we developed for the online brokerage firm addressed these needs by providing practical advice for moving funds into what would be safer investments, and by providing a forum where those nearing retirement could share ideas for coping with a less well-funded retirement. In the end, though, I realized these would only be relatively short-term solutions, especially if those betting on the economy making a full recovery soon are wrong.

Stories like those of James, Allison, and Sam are symptoms of an economic system that is beginning to become increasingly unpredictable. Some argue the ups and downs are a normal part of economic self-correction. Others say the current economic instability is a temporary setback due to irresponsible personal credit management, overly optimistic market speculation, corporate greed, or ineffective government regulation—all of which are, in theory, fixable. While each of these may play a role, economic instability cannot be blamed solely on them. According to economist Bernard Lietaer, these are epiphenomena traceable to a more fundamental problem.

Lietaer is one of the architects of the Euro currency and someone accustomed to looking at the broader economic picture. He sees our current economic system as unstable and believes collapse will be inevitable unless we change our monetary ecosystem. Lietaer's diagnosis is that the lack of diversity in the monetary ecosystem diminishes its resilience and sets up the conditions for collapse—much like a biological ecosystem becomes easily compromised when it becomes a monoculture. He notes that, "For the growing movement of communities, organizations and governments working to turn things around, the most daunting challenge often revolves around money: How do we find the money needed to transform our energy, food and health infrastructures, to eradicate unemployment and create green jobs, clean up the environment, and ensure that people have proper access to housing, education and meaningful work? Have you ever wondered why cash shortage so bottlenecks our best efforts and initiatives when we actually live in a world where there is neither a shortage of things needing to get done, nor a shortage of people wanting to do them?" [33] His recommendation is to create sets of complementary currencies that could facilitate exchange alongside the national currencies that already exist. You can watch Lietaer explaining his vision by going to the QR code on the left.

Lietaer video on complementary currencies

Whether or not Lietaer's diagnosis is correct, his questions do shine a light on one way in which the economic system has become complex. We have plenty of desire, and human and material resources to do much of what we might want to do to improve our lives, but doing what we want to improve our lives within the current economic system has become too complex. Allison and James may want to start their own business, but access to U.S. dollars is now cut off for them. How are they to untangle themselves from the debts and obligations they still have when job prospects are scarce and starting their own business requires capital they don't have access to?

Engaging an alternate currency or medium of exchange is one option, and many of these alternative currencies are becoming accessible online. Superfluid (https://thesuperfluid.com/) is one such service that facilitates community exchange of products and services through a virtual currency (called "Quids") so community members can keep track of trades. There are hundreds of other such mutual exchange services and complementary currencies in use around the world, including Local Exchange Trading Systems (LETS) and Time Dollars.

Complementary
currencies list

Employing alternate and complementary currencies may be useful and inject some resilience in the economic system—as well as give Allison and James some options—but that does not address the more fundamental question of how to create a worklife that is more satisfying and fulfilling. To do that, we have to consider what it means to earn a living.

Rethinking what it means to earn a living

Let's go back to Madison's story. The definition of work

that Madison's parents are buying into when they became puzzled by her choice to quit HDM is so pervasive you may not even question it. The assumption underlying much of the world's work ethic is that work is supposed to be a shorter-term sacrifice for the longer-term greater good—whether that good be earning money to feed your family or paying your dues at the bottom of the corporate ladder for the chance to attain a position of respect. The traditional mainstream payoff for working hard comes at retirement age, when presumably you have the funds saved and free time to enjoy doing whatever you like. The gap between input and outcome is wide. This kind of delayed gratification is deemed a noble sacrifice instead of what it often is—a way to let your 'happiness muscles' atrophy.

Attitudes about work are shifting among some toward working less, downsizing, and simplifying needs in order to take more time off and to keep overhead as low as possible. While downsizing, working less, and the voluntary simplicity movement does at least acknowledge that you need not postpone fully enjoying life until retirement, they often still rely on making a short-term sacrifice—that of income.

Erica, a friend of many years, opted out of the corporate life five years ago in order to travel the world and blog about her adventures. She either sold or put all her possessions in storage—except for a laptop, backpack and survival essentials—and hit the road for Asia. She has yet to return and at last chat has no plans on doing so. The upside of Erica's choice is freedom to travel and do what she wants. Some may see the potential downside being she has little room in her budget should an emergency come up, so that she may have to rely on staying with friends or housesitting engagements when she needs to return to the relatively expensive

States, and if she needs to replace her laptop, she must rely on donations. Erica holds the radical belief, however, that she will get what she needs when she needs it. She values living a life rich in experiences above all else. Experiences are her personal economic currency. Given her track record over the last five years, I have no doubt she has mastered the art of getting what she needs when she needs it and turning her thoughts into reality in a way many would consider magical.

For most of us who have not achieved this level of mastery, we consider working full time worth the sacrifice of freedom so we can enjoy the comforts of home and having the ability to fully care for ourselves in case of emergency. Until recently, only those with admirable drive, creativity, entrepreneurial determination, and a bit of luck could build a business that gave them the leverage to work less while retaining a sufficient income. This may be changing as the leverage that enables entrepreneurs to become successful becomes more easily accessible to the average individual. This change has everything to do with the convergence of social media and technology.

Several trends are making it possible for almost any reasonably intelligent worker to create a business of their own and reap the benefits of entrepreneurial leverage. These include access to investment funds through crowdsourced funding, access to specialists, the ability to outsource work online, and easy access to the tools of production with a minimal investment.

Let's take the entrepreneurial example of Scott. Scott got laid off from his job as a product manager for a medical insurance company. He sent out dozens of resumes, started networking and even got a couple of interviews lined up.

He still had quite a bit of time on his hands, so he started to think about how else he might fill the days. Just before he got laid off, he had purchased an iPad. He found himself using it more than expected, and soon noticed the device was starting to lose its look of newness. All the iPad cases he found online seemed absurdly expensive for boring cases, and none really showed any style. He figured if he could design and build an iPad case that *he* would like, others would like it as well, and he could sell it.

Scott had an idea of what he wanted—something cool-looking that would protect his iPad using as little material possible. He had no design experience, so he wrote a description of what he wanted and submitted it to a contest on 99Designs.com. The winning design would get a US$250 award. He ran the contest for ten days and received 32 mockups. Of the 32, five were close. He gave the five runner-ups constructive feedback and at the end of the ten days was able to select a winning design from among the five updated designs.

Scott paid the winner an additional US$100 to create a digital 3D model of his design. Once he had the model, Scott went to Shapeways.com, uploaded the model and paid US$50 to have the iPad protector prototype produced. The final product was better than he expected, but more importantly, Scott was inspired by the process of bringing his vision to life with little expertise or money of his own to invest. He wanted to make it easy for others to design their own custom iPad covers and buy them.

To get the business going, Scott would need to pay a designer to create a few mix-and-match templates that buyers could use to come up with a somewhat customized design. He would need to purchase a 3D printer—a machine that extrudes hard plastic in layers to form a three-dimensional object—and printing material that could manufacture the finished designs. He figured he could use a blogging platform like Blogger, Typepad or Wordpress to set up a web presence and he'd just need to hire a programmer to build the interface for customizing the iPad protector. He would link to Google Checkout to process orders. And he needed some funds for marketing. All in all, he estimated he could pull the whole thing off for US$3000—$3000 he did not have. Someone told him about Kickstarter—a funding platform to get non-charitable creative projects off the ground. Scott posted his project idea along with a tiered incentive for funding. Those who contributed $20 or less would get a thank you on the site and 15% off their first purchase. Contributors of $21-75 would get a founders' mention on the site and 15% off the first year of purchases. Contributors of $76+ would get all that plus one free custom case of their choosing.

Scott already had gathered a decent Twitter following of a little over 400 people by posting what he thought were funny quotes, a curated version of the news, and observations from his own quirky perspective. He asked his followers to support his Kickstarter campaign. Within a few days, he had pledges of US$2400. The remaining US$600 trickled in over the next couple of weeks. In less than three weeks, Scott had raised enough to start his business and put the job search on the back burner.

Fast forward three months later, between his blog posts being picked up, being retweeted on Twitter and the increase in his Facebook fan base, Scott is filling an average of 100 orders per week. The custom version of the iPad protector costs him about US$10 to make—including production, materials, marketing and other costs of running a business. He sells each one for US$50. At his current rate of production, he is grossing upwards of US$16,000 per month—more than he ever made working as a project manager. He plans on expanding to other customizable products using the same process, purchasing a few more 3D printers and ramping up production. His is still a one-man show. He currently puts in 25-35 hours a week max, and with more 3D printers he can increase his productivity exponentially without having to hire and manage employees. His next goal is to improve the efficiency of his process and rig the printers so he can feed them a design and operate them remotely. He can then work part time from anywhere while earning enough to do what he enjoys.

Back to our former architect friend: Madison is thinking along similar lines. The reason she got into architecture in the first place was to make cool things. She knows she has a great design eye and the training to pull off the creation of whatever she imagines. Madison wants to channel her talent into the furniture making business. She figures with an investment of about US$1500 in computer equipment and software and another US$1500 in 3D printing equipment and supplies that she sourced using IndieGoGo.com, a service like Kickstarter, she can start to build working prototypes. To sell them, she plans to partner with online retailers of home furnishings. Her process will take longer than Scott's, but she expects to sell her pieces for anywhere from US$200 to $2500 each. She thinks the prices are affordable enough for the average person who appreciates innovative design so anticipates getting as much business as she wants to take on.

Madison has one more selling point in her back pocket—a kind of radical idea. She is planning on piloting a program where people who purchase her furniture can send it back to her at any time to be melted down and recycled into a fresh piece of a similar size for about half the price of a new one. She envisions a world where nobody throws out furniture when it no longer fits, but rather takes the old piece and uses the material to create something that does. She is not sure how this will work out logistically, but she knows she will figure it out.

Each element enabling Madison and Scott to build a low-maintenance work life they enjoy is available in the marketplace today. The rise of social media and technology is providing inexpensive and easy access to expertise, information, production equipment and processes, and funding—if you have a worthwhile idea. Madison and Scott have abandoned the notion that they must sacrifice the bulk of their waking hours to work so they can build a nest egg for the future. They have the tools to create things they themselves enjoy, and ones they believe others will benefit from and enjoy, too. This "work" need not take up all of their time or require them to be in any particular location. The first question Scott asks himself is, "will this be fun?" If the answer is no, he does not do it. For Madison the question is more like, "does this make a good contribution to the world?" and if not, she knows it will not hold her interest long enough to make it real.

Making the best use of time

You might consider how work would change if you stopped asking yourself how to turn time into money and started asking yourself how to turn time into happiness or fulfillment instead. In times past, we used to have to all pitch in to a central organization where shared labor was necessary to produce products and services. Now that we have so many more tools and the material boundaries of production are going away, shared labor is becoming less of a need. The process is becoming decentralized, and participants in the economy can have more autonomy without having to sacrifice income. The barriers to education are beginning to erode as interest groups form around science and manufacturing among others. The lines between work and pleasure are fading as well. More than ever before, you have the opportunity to pursue what you are interested in and to turn that into your way of making a living.

If you are like most people, you may be unaccustomed to considering what you really want to create in the world. Some of us are like the dog kept for months on a six-foot leash staked to the ground who does not consider venturing beyond the six-foot diameter circle even when the leash is removed—until someone points out he's free to go further. Like this dog, many of us can benefit from someone pointing out the leash is gone. The availability of alternate currencies, fundraising platforms, at-home manufacturing machines like 3D printers, crowdsourced design, access to knowledge—the list goes on—perhaps gives you practical tools to start rethinking what it means to earn a living.

A new value for work

One more idea to chew on: Imagine we have improved just-in-time production to the point where we can get whatever we need and want easily, as posited in the earlier "Your Daily Life" chapter. The energy we formerly had to put into work and the production and consumption of goods is now largely freed up.

As old limits are lifted, there is often a period of anxiety as we become aware that our existing worldview was optional all along. During the dot-com boom, I saw many people who experienced sudden wealth freak out a little as they realized they could buy or do almost anything they imagined. There is something about the safety of limits that many of us find comforting. Conversely, there is something about the vista of unknown potential that many of us find disturbing. The impulse is to normalize our experience as quickly as possible using familiar tools. The point is to begin thinking about the impact of the rise of social media and technology now, just as it is beginning to ramp up. It is best to think about how you will make the transition before you find yourself in the middle of it. If the time comes when we can almost instantaneously manifest what we need to live, what new value might "work" have?

CHAPTER 6

The DIY Community: Creating the World You Want

Asking yourself, "What do I want?" for your community is no longer an act of idle daydreaming. It is a practical first step toward creating the world you want to live in.

Do-it-yourself citizen action

Residents of Kauai heard news that flood damage to roads in heavily touristed Polihale State Park would take the cash-strapped state of Hawaii *two years* and US$4 million to remedy, likely forcing several businesses to shut down. This would have devastated the local economy that relies on tourism for the majority of its income.

Eight days later the repairs were complete—for almost free. Did the Hawaiian government realize the potential impact on local business, get really clever about sourcing supplies, and kick the project into high gear? No. Residents whose livelihoods were affected by the damage volunteered their time and donated supplies to get the job done.[34] Welcome to do-it-yourself (DIY) civil works in action.

Unlike Kauai's case, DIY civil works projects need not be in response to an immediate crisis. A handful of citizens of Estonia were tired of seeing their forests used as a place to discard old furniture, obsolete machinery, used tires, and household trash. Most of their fellow citizens thought no more of using the forests as a dumping ground than US citizens thought of tossing trash out car windows along the highway prior to 1970's public awareness campaigns. Estonian project initiator, Toomas Trapido, asked a simple question: "What if we clean up the *whole* country?" Fellow collaborator and project visionary, Rainier Nolvak, upped the ante: "Stop. We will do it, but we will do it in just *one* day."

The team had no idea how to pull off the cleanup of an entire country, let alone in a single day, but they did not let that intimidate them. The *Let's Do It! My Estonia* project was born.[35] They embarked on a grassroots campaign to generate buzz and attract volunteers. Over 700 volunteers used mobile phones to geotag more than 10,000 trash sites, which were plotted using Google Earth to create a real-time virtual garbage map. With the visual of the garbage map in hand, they were able to demonstrate to the citizens of Estonia just how much garbage littered their forests. They engaged Estonian celebrities, business leaders, and volunteers in an awareness raising campaign to generate interest among citizens to participate in the one-day cleanup—with great success.

Two weeks prior to cleanup day, they had 10,000 volunteers registered, which was encouraging but not nearly enough to meet their goal of cleaning up all the garbage in the entire country. Then the campaign hit a tipping point. More than 40,000 people signed on in less than two weeks. On cleanup day over 50,000 volunteers showed up—an impressive 4% of the entire country's population. If the Estonian government were to have taken on the project, they estimated it would take three years and $22,500,000 Euros to complete. The Estonian citizens did it in five *hours* on a single day with approximately 2% of the budget estimated by the Estonian government. At the end of the day, the forests were totally cleaned of trash.

Let's Do It! My Estonia *project video*

The Hawaiian and Estonian projects each started with a simple question: "What do I want for my community?" In the case of the citizens of Kauai, the question was generated in response to a crisis. In the case of Estonia, the question came up to address an eyesore that had been bothering some citizens for a while.

You need not wait for a crisis or some scenario in your local environment that triggers a reaction, however. You can ask yourself, "What do I want for my community?" today. There are concrete ways to start creating the environment you want right now.

The DIY city

John Geraci founded DIYcity.org in October of 2008—a few months after I started research for this book. Geraci had a practical vision for participatory civic engagement by creating open-source tools that can draw upon the collective intelligence of city residents to find ways to create the city they want. I remember being inspired by his vision at the time, and expected it would be close to a decade before we would see anything close to a working model. Over the next year or so, DIYCity spawned local interest groups in fifty cities across the globe. Geraci initially set up DIYCity.org as a simple blog to spark a conversation about how we can make cities work better by combining physical space and the web. "It just steamrolled from there," he says, "pretty soon it caught fire and people started setting up their own groups and open-source projects."

Geraci acknowledges both the necessity and challenges associated with revisioning how cities work as technology and the web become an implicit part of their governance. "Our current cities are built of asphalt and brick, which doesn't take into account how large a part technology now plays in how they run," he notes. Unlike some of the other traditional industries like music and publishing whose foundations have been undermined by advances in technology and the web, he realizes that "traditional city planners are right to be a little concerned about where and how we integrate technology. If a new form of city planning doesn't work, we can't just get rid of cities." As he grappled with the practical reality of realizing his vision, Geraci has come to appreciate that there are "a lot of different pieces coming together when we think of the city of the future as an 'open city,' including open data, open source, open platform and an open contract process."

Open data is perhaps easiest; it's providing transparent access to data that exists. Open source development currently tends to work for producing low-level software solutions and is not yet that good at addressing more complex computing needs but that can be resolved by finding incentives for skilled developers to participate. Open platform includes applications and services that anyone can create, modify and use—these are already in place in some markets. Adopting a process of open contracting will be the next hurdle to overcome according to Geraci: "The last domino to fall that may cause a lot of disruption to the way cities currently run is moving toward an open contract process. The big roadblock to any civic hacker is getting a city contract; the cycle is so long. What we need is an open contracting system where the government puts out a spec describing what they need and people just build it. The city and its residents can then decide which solution they like best, and award the contract to that hacker/developer. If one solution gets sixty-five percent of the vote and another gets thirty-five percent, you could even award a percentage to each. Right now the way the government works is slow and doesn't encourage innovative solutions. If the Internet had worked this way, we'd still be using Alta Vista for search."[36]

San Francisco is one of the cities inspired by the idea of an open city, and is currently providing an open-source platform for local collaboration. In August of 2010 San Francisco announced the launch of Open311.org in collaboration with the city of Washington, D.C. Open311 is designed to be an application programming interface (API) that can be used, inter-modified, and improved upon by programmers and citizens in cities throughout the world. Initial goals are a relatively modest extension of current mechanisms that are in place for citizens of San Francisco to report if a building has been tagged with graffiti or if some other aspect of its public space needs attention—like a broken streetlight or a pothole. Instead of each person individually reporting something to be fixed to city officials—where it then enters a kind of "black box"—residents can take a photo of the graffiti and send a report to the common platform. Everyone can contribute to the platform—adding information to an existing report or creating a new one. Having all this information aggregated on a platform accessible to citizens and those responsible for fixing the problems alike makes it easier for everyone to see where the hotspots are, what has been fixed, and any projects that may be lagging.[37]

The potential of an open civic engagement platform goes well beyond fixing your local city potholes. Imagine thousands of projects like Kauai's Polihale State Park road repair and Estonia's forest trash cleanup all being fed into and coordinated on an interoperable platform. Instead of a series of one-off local community projects that start from scratch each time, you can generate shared learnings across similar projects around the world. You could learn which technologies or social incentives work best to reduce graffiti or perhaps make it part of community art like Mumbai's The Wall Project (www.thewallproject.com), ideas for reducing costs, tips on encouraging volunteer participation—the list goes on. The platform could be used to brainstorm fresh ideas, become inspired by the ideas of others, find new resources and monitor the health and well being of the local environments and their citizens.

Individual grassroots projects are nothing new, but this is the first time in history when it is possible to create a decentralized network of collaborative governance. On a small scale, leveraging social media and technology to improve collaboration and transparency makes sense. Could it work on a large scale to replace the bodies of political and corporate governance that rely on a top-down, centralized system of command and control? And is there any reason we would want to?

The challenge of centralized organizations

Centralized, hierarchical organizations have worked in some form or another for thousands of years. Top-down leadership with a clear chain of command can help members of the organization understand their role in the hierarchy, what the goals of the organization are, and the rules for participating. One of the challenges of centralized leadership within traditional organizations, however, is that the priority frequently becomes maintaining the structure of the organization above all else. Traditional governmental and nongovernmental organizations take effort to form, funds to operate, and protocols dedicated to maintaining their stability. They can be fraught with political infighting and jockeying for resources and power. Personal goals and desires are subordinate to those of the organization. Likewise, a sense of personal responsibility can erode among those who report up the chain of command, while those near the top of the chain who are presumably responsible are frequently unaware of what is happening on the ground. In short, the feedback loops meant to maintain the health of the organization can easily become disconnected. Often what the organization was set up to do in the first place comes second, while maintaining the organizational structure comes first.

The events leading up to the 2010 BP *Deepwater Horizon* oil spill, that ended up dumping 206 million gallons of crude in the Gulf of Mexico and killing countless birds and sea creatures, illuminate some of the challenges of a traditional company structure.[38] A large oil company like BP must rely on the soundness of its protocols, adherence to regulations, and an established chain of command to ensure it continues to extract crude oil from the earth and sea without initiating an environmental disaster. Following the outcome of the oil spill in the Gulf of Mexico, it became clear that critical points in the feedback loop had become disconnected—only after it was too late for crisis to be averted.

In a September 2010 report issued by BP summarizing their findings, BP executives concluded: "The team did not identify any single action or inaction that caused this accident. Rather, a complex and interlinked series of mechanical failures, human judgments, engineering design, operational implementation and team interfaces came together to allow the initiation and escalation of the accident."[39] Reading the report in BP's own words, you get the impression that dozens of employees and other participants in the drama played their narrowly-defined roles within the organization while putting on blinders to what must have been obvious signs that the system was breaking down on all fronts—mechanical, technical and operational. It is easy to conclude that people did notice, but either did not want to risk overstepping their bounds by escalating without a clear pathway for doing so—thereby jeopardizing their job—or perhaps they did speak up but did not have the authority to affect change. The BP report is deliberately vague on that point because the internal investigatory team says they did not have access to interview enough of the staff to form a clear conclusion.

The BP *Deepwater Horizon* story highlights one weakness in a traditional corporation: The motivation to participate tends to be the promise of keeping or improving your position in the organization—regardless of how much you are actually interested in what you are doing. The price of not playing the role you are given is the threat of losing that position. The result is that by participating in a large, centralized organization its members tend to get disconnected from their individual desires—like raising a red flag where they see something amiss—in favor of keeping the structure of the organization intact, along with their place in it.

"Because it's my neighborhood"

Even not-for-profit organizations formed expressly for the purpose of humanitarian aid are subject to the downsides of centralized leadership. Take the example of USAID's MarChE (Market Chain Enhancement Project, www.haiti-marche.org) project in Haiti. Even though the project was formed and funded to help improve the economic situation of Haiti's mango farmers, in the end it failed to deliver because of the limitations inherent in being a centralized organization. The story does have a happy ending, though a happy ending that had little to do with MarChE. First a little backstory:

Jean-Maurice Buteau, a Haitian mango exporter, gets most of his mangos from small farmers (www.mango-haiti.com/organic.htm). The problem with sourcing mangos for export from small farmers is they have few facilities to wash and store the mangos once harvested, so the mangos end up in piles under beds and randomly stacked while they await pickup. This means many get bruised and damaged, so relatively few are in good enough shape to be accepted for export to the United States.

Buteau wanted to build a center where small farmers could wash and store mangos, thereby keeping them in better shape so that more mangos would be suitable for export and the farmers would see a higher profit. He made his case and got support from some local NGOs. USAID MarChE, one of the local NGOs, received funding to help get the mango processing center off the ground.

One of MarChE's responsibilities was to get water to the center. The water source was 27 miles away, which required hiring people to dig up an existing pipeline and route the water to the center. MarChE was supposed to hire people to set this up. After many false starts, USAID cut MarChE's funding, leaving the center without an important resource to get the project going. When funding got pulled, the project was left without its most critical resource—water.

In the end, local residents did what they perhaps should have done in the first place. They did not look to a traditional governmental organization to fill their needs. Instead, the local community took it upon themselves to coordinate the resources to build the water conduit. Some community residents volunteered their time as manual laborers to dig up the 27 miles of pipeline, while others volunteered to feed the workers. When asked why he would do all this work for no pay, one volunteer worker simply said: "Because it's my neighborhood."[40]

The mango processing center is now well underway, due in large part to local people coming together to fill a need that is relevant to their lives—a need that could only have been an abstraction to traditional nongovernmental organizations. MarChE could not have pulled this off because its structure and protocols were not flexible enough to respond to the immediate need once funding was eliminated. The model it had to work with required funds to hire engineers and workers to build the water pipe to the center. With that gone, the project had no flexibility to arrive at a different solution.

At its heart, the very existence of an organization with a

set structure and centralized control creates a longer feedback loop between the purpose of the organization and meeting the needs of those it was created to serve. The best solution in the case of the Haitian farmers came from those unencumbered by a formal organization. The farmers themselves could focus their intention on getting the job done without worrying about following a pre-determined protocol.

Benefits of decentralization

One difference is that decentralized groups, in contrast to traditional organizations, can be dynamic. There is no centralized leadership that can take away the power of individual members should they not conform to conditions the leader may impose. Individuals can come together as long as makes sense and disband when the shared goals are complete or abandoned. The feedback loops are short and closely related to individual desires. If something is not working, there is an incentive to either fix it if it is worth fixing, or let it go if not. The incentive to participate becomes creating something you really want, while the disincentive becomes a possibility of not achieving your goals. Most importantly, the motivation to participate is *intrinsic* rather than *extrinsic*.

Since 1991, every year at the end of August a temporary city springs up in the middle of Nevada's Black Rock Desert. Black Rock City and the Burning Man festival (www.burningman.com) are a brilliant example of what happens when thousands of people, driven by an intrinsic motivation, coalesce to participate in creating community.

After years of saying "no, thank you" to friends' invitations to join them at Burning Man, I finally attended for the first time in 2005 and again in 2010. Despite reading up on reports from those who had been and seeing numerous pictures and videos, I was completely unprepared for the live experience of Black Rock City. Prior to attending, I chalked Burning Man up to being a giant rave held under harsh conditions. Year after year, I thanked friends for the invite, but said I would rather spend a week of free time hiking in the Yosemite backcountry or heading to Mexico for some sea kayaking. I didn't really understand why otherwise smart and responsible adults would choose to spend their free time for a week-long rave in the desert. I get it now.

The beating heart of Black Rock City is not a single visionary or charismatic leader. It is the collective energy of 50,000 people (give or take, depending on the year) who choose to actively participate in forming community *just because they want to.* And by participate, I do not mean simply gathering for a party, though there certainly is that aspect writ large. The sheer scale and multiplicity of voluntary creative contributions—from a full-on *Cirque du Soleil* style giant tent performance with flying acrobats, creatively dressed stilt walkers and magical sets to a 40-foot dragon that breathes fire as it moves along the desert floor, to elaborate nightclubs complete with huge art installations that pulse fire in time with the techno beat—is like nothing I have ever experienced elsewhere.

Aside from a small group of organizers, healthcare workers, and law enforcement representatives who help keep a city of this size loosely coordinated, every single structure, piece of art, performance—you name it—is created and freely contributed by those who volunteer. The incentive to participate is the desire to do so. If there exists a disincentive, it would be the possibility of not fulfilling a desire. Nobody is forcing anyone to come or participate.

Other than ice and beverages available for sale at Center Camp, no money is exchanged and bartering is discouraged. Burning Man officially runs on a gift economy where all that is given is given freely and all that is taken is taken with gratitude, nothing more. This ethos creates an atmosphere that pulses with positive energy—which to me is truly the beating heart of Burning Man. Once the week is over, Black Rock City—including its airport, buildings, roads, and works of art—vanishes without a trace as its residents voluntarily pack out everything they brought with them. To me, the possibility of generating this kind of positive energy is one of the biggest potential benefits of starting to explore decentralized forms of governance.

Decentralized governance may sound feasible when relatively few people are involved, as with the Haiti community—or when the task is clearly defined and can be accomplished within a set timeframe, like the Estonia project—or when it is just for the joy of it, like Burning Man. When ongoing governance of long-term goals and large numbers of people is required, however, creating a decentralized system of governance driven by voluntary participation probably seems hopelessly idealistic to many of you. Inevitably, there will be those who will want to game the system for their own advantage. And getting large numbers of people to agree is a notoriously difficult task, especially when stakes are high. So, how can decentralized governance possibly work?

Coordinating the masses

Building consensus and getting everyone to play nicely together has always relied on persuasion of one kind or another. Persuading people within a decentralized system and on a large scale without relying on a charismatic leader or the threat of, say, legal action may seem like an impossible task. Fortunately, online platforms for decentralized governance make it possible not only to encourage transparency and participation, but they can also address a key problem many governing organizations face. Let's call it the "Town Hall" syndrome.

The Town Hall syndrome is found, for instance, where citizens are invited to have their voices heard equally by government representatives. The objective is to allow individuals a voice in the decision-making process and to build consensus. In San Francisco, with its legendary subset of highly vocal residents, you will often see people with the strongest opinions and loudest voices dominating the public discourse. Speaking over lunch with a few city planners recently, I learned this is one of the challenges of governance they spend a lot of time addressing. Citizens who are loud and opinionated tend to drive away more reasonable residents who are interested in coming to a solution. City planning consultant, Jeffery Tumlin, observes that "in larger cities the conventionally run public processes tend to get dominated by professional media tenders or crazy people—the extremists; in San Francisco that means identity politics."

Successful governance means putting a lot of time into encouraging more reasonable people to stay in the process while keeping the most vociferous from taking over the agenda, according to Tumlin. One of Tumlin's strategies is to rely on face-to-face interactions that can build coalitions of trust among city residents to strengthen the voice of the collective. "There are aspects of the social contract that don't come into play when we're not in the same room," he explains, "you have to build trust. As human beings we get an enormous amount of information through eye contact. The human eye is unique among animals in that we have the greatest contrast between the colored iris and whites of our eyes. Getting information about another person's trustworthiness or lack thereof via the expressiveness of their eyes is hardwired into our brains."

Tumlin acknowledges that online platforms have the advantage of giving a voice to, say, those who like to ponder issues before responding or those with disabilities make it difficult to attend in-person meetings. However, he believes that "it can be trickier to create certain kinds of trust in a negotiation unless you have face-to-face contact," especially where complex tradeoffs are involved. "Particularly in city planning—in the planning world there are so many technical issues at play that are beyond the average person's ability to grasp easily. Having negotiation happen in real time can be important to get people to agree to a whole stack of conditions." He is skeptical, for one, that a technology platform can be as practical or effective in creating consensus as in-person persuasion in cases where a populist majority rule does not apply, such as in the case of minority rights. If, say, the majority of city dwellers see bicycles as a traffic nuisance, the likelihood of passing a proposal to create bike lanes for bicyclists, who are in the minority, is slim—despite the potential longer term benefits of less carbon emissions, a healthier population and less congestion on the street.[41] Persuading the majority to consider the positive long-term potential, even if it means sharing the road with bicyclists, is perhaps more likely if the majority trusts proponents of the bicycle lane proposal. Tumlin believes that building this trust is better accomplished in person.

You may agree that in-person persuasion is an effective means for building consensus. In the end, however, those with personal charisma are not unlike the opinionated vociferous citizen of the town hall. The difference is just that a charismatic leader's methods perhaps show more skill than the town hall extremist. Consensus builders may take a "carrot" approach to promote their agenda, while the citizen who secures a bully pulpit may take a "stick" approach. In either case, the democratic process may be left somewhat compromised *especially* because it relies on personal charisma. Using a decentralized platform along with persuasive technologies holds a better chance to take us a step closer to a truly democratic process, even where persuasion is called for. Technology solutions can still incorporate both "carrot" and "stick" tactics, but those tactics can be deployed from a neutral position. Expert representatives, too, have the opportunity to build consensus online through persuasive arguments moreso than personal charm.

Let's say you and your neighbors want to improve awareness of energy consumption throughout your neighborhood. You are persuaded by arguments that say energy consumption monitoring is a first step toward better energy management overall. Your homes are networked online so energy consumption is monitored for each home. The total of energy available for the entire neighborhood is fixed. Each day you can look at a monitor that tells you how much reserve energy you have available for your use as well as that available for your neighbors. You see that you have reached your limit toward the end of the day, but others are well below their daily, allotted usage. You decide it will be okay to draw more power tonight than you are allotted for that day, because the risk of the whole neighborhood going over the limit is low. The next night your neighbor might find themselves in the same position and you would be fine with them taking a little more than their share that day. In the end, you are participating in a technologically-mediated reciprocity that can work for everyone.

Now let's say it's time for the World Cup finals and an unusually hot day to boot. Everyone's energy consumption is starting to max out. Instead of accepting that you will need to turn off the TV and air conditioner, you decide to leave everything on and deal with the consequences later. The consequence is not that your power will be cut off; it is that a random neighbor's lawn will not be watered the following day. Did I mention the weather is unusually hot? On Monday, your neighbor's lawn is fried to a crisp because you decided to use more power than the system had capacity for. In this case, you know your actions did damage to someone else's property so you have incentive to play by the rules or else trip off a negative feedback loop where nobody plays by the rules and everyone suffers.

NaturalFuse is a proof-of-concept project that uses just this kind of idea to build awareness of energy consumption. NaturalFuse is based on the idea that the carbon offset green plants provide by absorbing carbon dioxide and converting it into oxygen can be quantified, as can the carbon dioxide produced by using electricity. The simplest NaturalFuse system includes a

NaturalFuse website

single plant, a water delivery system, and a small 9-volt electrical device like a lamp. The system has three settings: off, selfless, and selfish. The "off" setting operates the automatic watering system only enough to keep the plant alive. Other than that, the plant is absorbing carbon thereby creating a carbon credit for future use. On the "selfless" setting, the lamp will only light up as long as the plant has absorbed more carbon dioxide than the automatic watering system and the lamp have produced—maybe ten minutes per day. On the "selfish" setting, the plant will continue to be watered and the lamp will continue to light up until the carbon surplus is used up. Once the carbon surplus is gone, the light will remain off and a "fuse kill" will be triggered, meaning the plant will no longer be watered—leading to its eventual death.

The real power of this concept comes when many plants are networked together in several homes. The advantage of the networked NaturalFuses is that not all appliances are being used at the same time, so in selfless mode a single appliance can be turned on without running a carbon deficit. If you would like to turn on your lamp, the aggregate carbon offset of all the plants means that decision should have no negative impact on the system as a whole. Say you really want light, but in selfless mode there isn't enough surplus to power your lamp. In that case you would have the option of switching to selfish mode. The downside of switching to selfish mode is once maximum capacity has been reached, not only does it trigger the fuse kill cutting off the plant's water supply, it also triggers a system breakdown. Once a system breakdown occurs, vinegar is dumped into the soil of a random plant networked within the system, quickly killing it. This will not be one of your plants, however, but somebody else's. The entire system is networked to a central dashboard on NaturalFuse.org, so you can see how your choices are affecting the system as a whole—positively or negatively. Moreover, when your actions have a negative impact on the system, you can see that there are real consequences of your actions and see how those consequences have a direct impact on real people.[42]

NaturalFuse makes one of the biggest blind spots we

have as human beings tangible: understanding indirect consequences of our choices and persuading us to play fair. If the near-term payoff of a decision is tangible and positive, and the consequences of that decision are longer term or do not impact us directly, our bias is almost always in favor of choosing the near-term payoff over long-term consequences. Overcoming bias relies on closing that feedback loop. Likewise, reinforcing that bias relies on keeping that feedback loop open.

This is just one example of how social media and technology can be used as both a carrot and a stick to reinforce consensus without relying on the charms of a local organizer or the bullying of a complaining neighbor. And it makes it difficult for individuals to game the system for their own advantage. That said, there will never be a 100% effective technology solution to generate cooperation and fair play. As long as there are people intent on subverting the system, the system will be compromised. Perhaps for those on the fence, in-person persuasion is an effective way to get them to cooperate, whereas they would not be compelled by rational arguments or by experiencing negative consequences doled out by a neutral platform. In the end, participating in a decentralized system of governance and civic action requires *wanting* to embrace an ethos driven by creativity and collaboration along with personal and shared responsibility. At present, this DIY ethos (singular and plural) can and does work in many diverse arenas.

The DIY ethos

Tapping into the imagination and being inspired to create may be the single most important driver of the DIY ethos. Carlos Owens exemplifies the power of being inspired to create. Carlos lives in Wasilla, Alaska. He is a former U.S. Army mechanic who now makes his living working the oil fields seasonally. During the off-season, 33-year-old Carlos spends his time figuring out how to make cool stuff like Big Red—an 18-foot tall Transformer-looking "mecha" exoskeleton that he can climb inside and walk around in and that shoots flames from its arms.

He has not yet built a model that can actually transform into a vehicle or anything, but the exoskeleton is impressive enough. Big Red took Carlos two years and US$25K to create—that is, after trying out models since he was 19 years old. When Big Red was finished, he says he was able to take it for a short walk. I could not find any video of that moment, so we will just have to take his word for it. Today Big Red stands suspended in a scaffold in Carlos' back yard, enduring the Alaskan weather. Carlos says it needs work, but he is now spending his time on what he sees as more practical projects: a hovercraft and a flying bike.[43]

Thirteen-year-old Hibiki Kono is another example of this kind of creative urge. Hibiki wanted to scale walls like his hero Spiderman. "I used to dress up like Spiderman when I was younger and I loved all the films," he told a *Sun UK* reporter. If a vacuum cleaner tube could stick to a surface using suction, why couldn't he use the same technology to stick to a vertical surface? Hibiki told his teacher about the idea. His teacher was not so sure the idea would work: "When he came to me with the idea at first I had my doubts," he said.[44] Equipped with two Tesco vacuum cleaners he purchased for less than US$50, a couple of suction pads and straps to attach to his feet, Hibiki can

Hibiki Kono climbs a
brick wall video

successfully climb a sheer brick wall.

Although others have demonstrated similar uses for vacuums in the past, there does not seem to be evidence Hibiki knew of earlier efforts. "He developed it himself," his teacher says, "which is amazing for someone his age."

Carlos and Hibiki's projects show an inspiring level of creativity, resourcefulness, and determination. They are both shining examples of individuals driven by a do-it-yourself (DIY) ethos. Inventors who combine commonly available tools and materials with elbow grease and a little imagination are, however, nothing new. There have always been a handful of creative souls with visions that time and determination make real. Until recently, however, DIY has been put in the category of "hobby" or something borne out of necessity where resources are lacking and ingenuity is up to the task.

With the advent of technology that puts once professional-quality manufacturing tools in the hands of individuals, and with plummeting costs of what was once sophisticated equipment only a large company could afford, DIY is expanding from the realm of hobby and "mother of invention" necessity into the realm of practical small-scale manufacturing. Only a few years ago, setting up a functional manufacturing operation would easily have cost over US$100K. Today, you can set one up for less than US$6K.

Not only are the DIY tools more accessible than ever before, but opportunities to collaborate with like-minded others are expanding with the rise of maker/hacker groups. Maker/hacker groups range from super-niche interest groups like The Open Organization of Lockpickers (www.toool.us) that specialize in learning how to pick all manner of locks and share that knowledge with others, to large organizations like Bucketworks (www.bucketworks.org), which makes its 25,000 square feet of space available for collaboration and knowledge sharing among self-styled creators for a nominal fee.

Make magazine (www.makezine.com) is a print and online resource dedicated to sharing how-tos among the DIY crowd. During a visit to the Bay Area Maker Faire in San Mateo, I spoke with makers who dedicated their time to creating everything from whimsical mechanical art pieces to builders of a working space shuttle named "Sensational Susan" that had purportedly passed early trial flights and, pending funding, would be able to take passengers into space within a few years.

Bre Pettis is a pioneer in the DIY Maker space. In the last five years, the "amount of tutorials on the web has grown— I'm not even exaggerating—five million times" according to Pettis, "now you can make anything with modular electronics, the Internet, open source code and a MakerBot" – an open source robot that uses an ABS plastic extrusion and input from a digital design to print three-dimensional objects. The MakerBot costs between US$750 - $1000 depending on what you need and can print almost any object up to 4"x4"x6" in size. Pettis estimates that with less than a US$6000 investment you can set up your own home manu-

facturing operation that can make just about anything you would like. If you are interested in setting up a manufacturing operation of your own, scan the QR tag for his recommended list of items and where to get them. If you want to learn what the maker com-

DIY FabLab items

munity is creating, Pettis recommends checking out Thingiverse.com: "I like it when people make practical things like bottle openers, doorknobs, and shower curtain rings, but it's also very popular to upload fun toys and art like toy cars, mechanical prototypes and puzzles," he says.[45] The RepRap.org community site provides plans and parts for a 3D printer that works in a similar way to MakerBot for approximately US$650. RepRap is open source as well.

The DIY ethos is extending far beyond manufacturing. Science, the practice of which formerly depended on trained specialists and often expensive, sophisticated equipment, is beginning to undergo the same transformation. Want to put your own satellite into orbit? Not a problem with a US$8,000 DIY satellite kit from Interorbital Systems. These personal satellites orbit at 192 miles above the Earth's surface for a few months before falling back toward Earth and being burned up upon reentering the atmosphere. While in orbit, self-fashioned rocket scientists can pick up their satellite's signal using a small handheld receiver.[46] Interested in DIY nanotech research? You can build your own scanning tunneling microscope.[47] Do you want to make your own robot? Knowledge sharing communities like LetsMakeRobots.com is the place to get ideas for the kind of robot you would like to create as well as practical advice on tools, components, and techniques that have (and have not) worked for your fellow robot builders.

Even biological research has entered the realm of DIY. In 2009, Eri Gentry took advantage of the rash of biotech companies going out of business to pick up US$1 million worth of lab equipment for about US$30,000, which she then installed in her garage and invited other self-styled citizen scientists over to roll up their sleeves and start experimenting. Her goal is to create a space where she and her fellow biology hackers can congregate and build it into a "ground zero" for the DIY scientist movement: "From Bio-Art to BioFuels," her site reads, "the wave of next generation biotech applications is set to transform our culture and economy." [48]

Manufacturing and science are perhaps the two most surprising arenas in which average citizens are gaining more direct access and control, whereas the recording and publishing industries have been increasingly DIY for a long time.

The recording and publishing industries were among the first to see the erosion of their roles in the production, marketing and distribution of music, newspapers and books. There are few good reasons to engage a traditional middleman when so many effective and inexpensive tools are available to produce, distribute, and market a book or collection of music. Traditional book publishers, for instance, still can have a role in curating niche audiences and providing access to a skilled network of editors, reviewers, and subject matter experts. Their value to the writer then becomes one of having a ready-made network for the writer to access. With the costs of production and marketing so low, the trend toward publishers putting the responsibility of marketing onto the shoulders of the writer, and the marketing and distribution channels no longer owned by the publishers (with minor exceptions), it is not a good deal for authors to accept the 7.5 percent of each book sale in exchange for a typical contract with a New York publishing house—unless the publisher is adding value in some other way.

The social trend toward a DIY ethos, enhanced information sharing through open source collaboration, and the increasing accessibility of educational and material resources all remove constraints that formerly would have prohibited a decentralized, DIY system to emerge from its hobby-ist roots. Some concerns have been raised, however, about putting sophisticated equipment in the hands of self-styled citizen scientists and manufacturers whose good intentions could lead to unintended consequences. It is true there are risks and potential unintended consequences, but the same can be said of trained scientists and manufacturers who participated in the creation of the petrochemical industry and big agribusiness. The BP example is one case in point.

Neither the centralized, hierarchical organization nor the decentralized, dynamic one is necessarily better than the other at mitigating risks and avoiding unintentional consequences. Expertise and specialized talents can be nurtured in a DIY environment, just as they can by having individuals complete traditional courses leading to academic and professional credentials. Guidelines and standards for excellence can be set in a DIY environment and reinforced by peers, just as regulations can be set and enforced by the leadership of a traditional agency. The trend toward DIY manufacturing, production, and science spaces that encourage open collaboration and information sharing, provide a dynamic environment where the barriers to creation are lowered due to the lower cost of trial, error, and the occasional success. Makers and hackers participate as long as they want to and are free to go at any time they choose. The lack of intermediaries, rise in inexpensive computing power, and DIY tools like 3D printers, make production and experimentation more playful, game-like. In this way it provides fertile ground for innovation that can be replicated in a more rigid and structured environment only with great difficulty.

Bringing together and focusing of human intent is half

of the equation. The other half is enabling the physical environment to participate in the communication. Responsive environments not only mean we can all make smarter choices about how to shape them, but they can be enlisted to change our behaviors as well. The example of NaturalFuse is a small proof-of-concept example, but much larger projects to bring the environment into the social network are in the works.

CHAPTER 7

Living in One Smart Universe

*People, products, plants, seas, bridges,
roadways, animals, satellites, and even
planets now have the capacity to contribute
information to a single universal "brain."
As these connections become ever more
massively networked, our power to make
smarter choices is amplified.*

Xavier Ousa, a 70-year-old rice farmer, tends four acres of fields of the *Pokkali* variety in Kerala, India. Xavier's farming practices are little changed from those of a long line of predecessors. Over the centuries, a symbiotic relationship between rice, people, and the local prawns arose in the brackish waters of the Pokkali fields—a system so smart and efficient, no proposed change has been able to improve it. Each year, prawn larvae are drawn from the Arabian Sea to feed on the nutrient-rich Pokkali field waters. On nights with a high tide, rice farmers put a burning light near the sluice gate to the fields, which attracts the baby prawns to enter. The prawns are caught on the field side of the gate where they live and feed in the fields for several months until they reach full size. After they reach adulthood, the farmers place conical nets at the gate. As water flows out to sea during low tide, the adult prawns are trapped in the net, providing food for the farmers. The bi-products of the prawns' lifecycle in the waterlogged fields create fertile soil for the rice to grow. After the rice is harvested, the stalks decay, which in turn creates nutrient-rich soil for the next generation of prawns to eat. There is no need for fertilizers or pesticides in this self-sustaining, intelligent ecosystem. Ousa describes the annual process:

"After all the prawns are caught by the end of April, water level is kept to the minimum using the *thoompu* (sluice gate) contraption, and earthen mounds of one meter base and 50 cm height are formed all across the field. In June, after the southwest monsoon brings in a few good showers, germinated Pokkali seeds are sown on the flattened tops of the mounds. In a month, the mounds are dismantled and the seedlings in clefts are dispersed around the flattened mounds.

"Since the tidal flows make the fields highly fertile, no manure or fertilizer need to be applied; the seedlings just grow the natural way. In order to survive in the waterlogged field, the rice plants grow up to two meters. But, as they mature, they bend over and collapse with only the panicles standing upright. Harvesting takes place by end-October. Only the panicles are cut and the rest of the stalks are left to decay in the water, which in time become feed for the prawns that start arriving in November-December. Then, the second phase of the Pokkali farming, the prawn filtration, begins."[49]

This ecosystem, unfortunately, is breaking down as one of the links in its feedback loop—human harvesters—either stop working when they receive social welfare assistance or elect to go into other, less labor-intensive careers.[50] C Bose, a fellow Pokkali farmer, complains, "These four women have been on the job for seven days. It's very hard to find farm hands for harvesting, these days." The work used to be a communal effort marked by annual ritual. Now that there are other options for work and since rice harvesting pays so poorly, one former farmer asks "Who will be interested in the work if they don't get cash?" and adds, "All my siblings have also quit the farms." As a result, much of the rice is not cultivated, which creates a poorer environment for the prawn larvae to grow, which makes it more difficult to grow rice successfully while adult prawn supplies dwindle—all in a downward spiral. The intelligence of the system that grew over centuries has now, apparently, run its course.

The Pokkali story is likely familiar to you if you pay attention to media reports on climate change or diminishing natural resources. Biological ecosystems have long feedback loops that do not tend to adjust well to quick changes currently being brought on by technology. Technology that allows this generation of Pokkali farmers to leave the family business in favor of a more lucrative living elsewhere sounds the death knell for the rice farms. With this single element abruptly taken out of the feedback loop (compared with the centuries of Pokkali rice cultivation), the fields are experiencing rapid decline. In this case, critical elements that made the Pokkali farm system work are missing and the system has yet to adjust, if it ever will.

What we have here is a failure to communicate

We are seeing these kinds of ecological breakdowns happening in every arena from the depopulation of sea life and giant, swirling patches of garbage in our oceans to glaciers melting in the Antarctic. The list of proximate causes can be debated endlessly. However, there is one fundamental cause they all share: Biological systems and technological systems do not as yet communicate with one another very well. While it may feel temporarily satisfying to point a finger at individuals who choose to eat endangered species of tuna, use plastic bags and drive SUVs, or at companies that drag heavy trawlers across delicate coral reefs on the ocean floor to catch as much fish as quickly as possible, or pump toxic fumes into the air as a byproduct of the manufacturing process, the reality is these are just epiphenomena traceable to a cavalcade of complexly-related decisions that came before.

The unfortunate truth is these decisions are often driven by a legitimate desire to improve quality of life for human beings rather than by a desire to intentionally do harm. The unintended negative consequences are frequently the result of human ignorance about everything from the importance of the role of corals on the sea floor for maintaining a healthy and sustainable ocean, to why it is unwise to depend on petroleum as a source for everything from the fuel that powers our engines to fertilizer for our crops.

This is not to say that greed and selfishness do not play a role. The reason we have an oil-based economy is arguably traceable to a self-serving decision by late 19th-century oil baron John D. Rockefeller to maximize Standard Oil's profits by strong-arming competitive products and technologies off the market. Once oil gained a cultural foothold, finding as many uses for it as possible to maximize profit makes economic sense from the perspective of the company and its shareholders. That said, even if a man like Rockefeller believed he was ultimately doing a good thing by making oil widely and cheaply available, he could not possibly have conceived that his decision would bear so much of the responsibility for the swirling mass of plastics circulating in the Pacific Gyre and the disappearance of Antarctic glaciers a hundred years later. Simply put, we as human beings are just not that smart. And up until now, the communication between biological and technological systems has been very poor—but that may be starting to change. We need a smarter system of communication.

Creating a smarter system of communication

If a better global communication system were in place a hundred years ago, we might have become conscious of the likely effect of decades of burning fossil fuels before we started seeing pictures of receding ice caps and melting glaciers. Without such a system in place, we can only resort to changing behaviors that stop contributing to global warming. Reversing the impact carbon emissions have already had, however, will now likely take a very long time. The biological ecosystem does not change course very quickly, so it is helpful to have tools that give us a better understanding of the potential impact of our choices. If we can get these ecosystems talking with one another, we may have an opportunity to improve our lives in ways that you might not expect.

It might surprise you to learn that such a communication system is already beginning to take shape. People, products, plants, seas, bridges, roadways, animals, and even planets are being tagged and connected online in what is being called the "Internet of Things." The Internet of Things is taking steps toward creating a single global "brain" of interconnected people, places, and objects. Parts of this brain already exist and can take in data, process that data into usable information, translate that information into action, retain memories of all this information and activity, and even, in some cases, actually learn from them. Depending on which expert you ask and how robustly they want to define what constitutes a global brain, we may see it come to full fruition in our lifetimes or never. I, however, will leave prognostication to the futurists and share what is already happening today.

Giving things and places memories

It may seem odd to say that inanimate objects can have memories. Perhaps it is more accurate to say that things can have memories *attached* to them that can both be added to and can be retained throughout their existence. That said, why might we want to?

Think of all the stuff you have owned or used in your lifetime—shoes, toothbrushes, books, cars, musical instruments, computers, works of art, photos. Currently we do not have much intelligence about where all this stuff comes from and where it goes. With routine tagging of items and connecting their cradle-to-grave lifecycle history to the Internet, we have the chance to become aware of how we use our stuff and what happens to it after we are through. One reason for connecting all this data is to give us an opportunity to make smarter choices about consumption and destruction of goods.

Products are among the first items beginning to be tagged with sensors during the manufacturing process so they can be given "memories" of their lifecycle. Walmart already includes radio-frequency identification (RFID) tags in jeans and underwear so they can be tracked. Tagging currently helps with inventory management, according to Walmart executives. This is only the beginning, however, according to Bill Hardgrave who leads the Walmart-funded RFID Research Center at the University of Arkansas: "We are going to see contactless checkouts with mobile phones or kiosks, and we will see new ways to interact, such as being able to find out whether other sizes and colors are available while trying something on in a dressing room. That is where the magic is going to happen."[51] It is not a big leap from embedding a RFID tag that connects your jeans to an in-store tracking and recording network to one that persists outside the store and records your jeans from the time you take them off the rack to its final end—hopefully to be recycled.

The RFID tag is a relatively simple tracking device that follows a product through space and time. With technologies like the QR tags you see throughout this book, or barcodes that can be modified and added to throughout the life of a product, the history of products can be embedded as "memories" to augment their spatio-temporal reality. Like the memory trigger embedded in the taste of madeleines described by Proust in *Remembrance of Things Past*, memories embedded in products can become multivalent through barcodes and tags that add a layer of information onto physical objects (augmented reality tags). As your favorite jeans live with you, you could use a QR tag or barcode embedded in the waistband to store video of you wearing them on vacation along with a favorite story. Whomever the jeans are handed down to will have access to your jeans' storied memories and have the opportunity to add their own, which you would in turn have access to. While adding stories to your jeans might be a novelty that isn't that compelling to you, if it were a treasured antique whose provenance is valuable in itself, or an item that has sentimental value you would like to ensure remains part of its future, the story might be different.

Giving objects memories then can serve both a practical and an educational purpose. All of us are largely unconscious of the numbers and disposition of things that we once used or owned. Have you ever moved and discovered objects you had long since forgotten you had? Or have you ever received a box of old photos containing an image of a favorite childhood toy that you hadn't thought about for years? As we enable objects with memories we can become more conscious of how we are using the time and material resources that make up the things around us.

Places can have memories attached too, by using much of the same tagging and tracking technology as that applied to things. You may have seen QR tags similar to those in this book in San Francisco or London. Hold your smartphone up to the tag by, say, a London Tube station, and you may see a popup text describing when it was built. More impressively perhaps are augmented reality applications that you download to your smartphone. Simply point your phone's camera toward an historical building or site and the applica-

Civil War augmented
reality video

tion recognizes that site through a combination of global positioning technology (GPS) and image recognition software. Up pops a visual layer over the physical site that may contain anything from a text description of the site to an image of what that site looked like a hundred years ago.

One such augmented reality application

in development is aimed at sharing the history of the U.S. Civil War. Visitors to Civil War battle sites like the Battle of Gettysburg in Pennsylvania may see the battle site as it appeared at the battle's end or a written history of the event. Another is NearestWiki that gives you an overlay of information and images as you find your way through cities including Rome and London. Conduct a search on "augmented reality" from within the iTunes application store and you will see dozens of examples of how augmented reality can be used. The opportunities for education and entertainment are limited only by the imagination.

NearestWiki website

Giving both places and things memories is an intriguing but small example of the potential for building one smart universe. Things begin to get a lot more interesting as objects and places are enabled with the power to communicate with us and with each other by being connected to the nascent global brain.

Here come the communicants

As sensors that transmit data from things to the Internet become smaller and more intelligent, as computing power becomes more powerful and less expensive—and as the devices that connect them all become less obtrusive and more instantaneous—communication among all the members of the network begins to approach frictionlessness. Everything and every activity have the potential for becoming part of a seamless network. This massively networked set of feedback loops promises to alter the entire ecosystem at an increasingly rapid pace.

One key implication of creating a global brain and better system of communication as just discussed is that we can make smarter decisions about resource usage and caring for our planet. There is another implication that may not have occurred to you: by participating in this network of communicants, we gain more power to create the world we want—and for a reason you might not at first suspect—but we'll get to that in a moment. First, what are the communicants that are beginning to join the social network?

Recall that the chapter "Your Massively Networked Self" covered some of the ways we as human beings are joining the network. "Your Daily Life" spoke to the parts of your home and the things that surround you that are being networked. "Rethinking Your Worklife" and "The DIY Community" shared how the social sphere is becoming part of this network. The environment that surrounds us— everything from oceans to cities to the planets in our solar system—is also becoming connected and joining among the communicants in this emerging global brain.

In Cape Cod, for instance, sailors heading out of Green Pond or Ockway Bay used to have to rely on National Oceanic and Atmospheric Administration (NOAA) tide charts to figure out whether they would have enough bottom clearance to get from Ockway Bay to the sea. Unfortunately, NOAA charts are frequently off just enough to let your sailboat bottom out when you thought you would have plenty of clearance. Then the Cape got itself hooked to the Internet.

Cape resident, Robert Mawrey, now knows precisely what sea levels are. An ultrasonic sensor bounces sound waves off the surface of the sea to gather real-time data about sea levels. The data from the ultrasonic sensor is relayed to a radio transmitter, which sends that data to a shore receiver. The receiver on shore uses an ioBridge transmitter to route that information to servers in the computing cloud, where anybody can access up-to-the-minute data on actual sea levels by going online and checking the ioBridge site. The ioBridge folks are hoping to evolve a worldwide network of sea level monitors that generate feeds that can, in turn, be translated into everything from maps to text messages to Twitter feeds.[52]

Current sea levels website

Cape Cod is only one element of the natural world that has a Twitter feed. Your houseplants can now tweet you when they need attention. Let's say you check your Twitter feed after a few days and see the following message from @spiderplant2: "You over-watered me" *three days ago.* You gave your houseplants Twitter accounts as part of the Bontanicalls (www.botanicalls.com) set-up. The backend is a simple moisture sensor embedded in the potting soil. Three days before that, you received a "please water me" message and did just that. Now @spiderplant2 is letting you know you didn't follow its instructions quite right. In the midst of Twitter messages from people you are used to communicating with, it may be difficult not to start thinking of your spider plant as one of them.

Cities are getting wired as well. Technology giant Cisco Systems is likely betting that this will translate into big bucks as they sell the sensors, connectors, and other technology that make wired cities possible. Speaking of plans to create dozens of new cities from the ground up to meet growing demand in Asia, Wim Elfrink, Chief Globalization Officer for Cisco in Bangalore, giddily predicts "Everything will be connected—buildings, cars, energy—everything. This is the tipping point. When we start building cities with technology in the infrastructure, it's beyond my imagination what that will enable."[53] IBM is another company on the forefront of facilitating the wiring of cities. Singapore used IBM's expertise and technology to create a smarter transportation system. Singapore's smart rapid transit system, SMRT, includes a card its residents use to ride all public transit, middleware that captures and connects all this rider data, and an analytics package that crunches all the data into information that modifies transport schedules and fares to optimize the system as a whole.

The energy grid is one more arena where smart technologies can improve the intelligence of the system. Utility supplier Xcel Energy helped install one of the first such systems in Boulder, Colorado, SmartGridCity™. Xcel executive Ray Gogel explains: "We like to think of Smart Grid as bringing the world of Thomas Edison together with the world of Bill Gates." Xcel Energy put solar panels on the one of the first homes to be outfitted with Smart Grid technology, and added a smart meter and a hybrid electric car to the system, which were all then wired to the Internet. Val Peterson, the homeowner, is delighted: "I pretty much get on my computer, tell my house and my car what to do and then I walk away," she said. "My solar panels are talking to my house, are talking to my car, are talking to my house. It's a beautiful system." It has worked so well for the Petersons that at times the system is generating more power than they use. Any extra goes toward charging the batteries in their car and backup power storage that will last them for a couple of days.[54]

Wiring seas, plants, homes, cities and the like shows promise for improving intelligence of local environments. Although these are impressive projects in themselves, they cannot become part of *one* smart universe until they are all connected within a universal network. HP's Central Nervous System for the Earth (CeNSE) project has an ambitious vision of creating a network "consisting of a trillion nanoscale sensors and actuators embedded in the environment and connected via an array of networks with computing systems, software and services to exchange their information among analysis engines, storage systems and end users" according to their site. The team expects this environmental monitoring to enable a new level of awareness and revolutionize communication between objects and people.[55] NASA and Cisco are teaming up on a similarly ambitious project to create a global nervous system they are calling Planetary Skin. The Planetary Skin project (www.planetaryskin.org) aims to integrate land-, sea-, air- and space-based sensors, to gather intelligence that can help mitigate the effects of global climate change by monitoring carbon emissions among other data.[56]

One of the most ambitious projects to expand the scope of this universal network is to wire outer space. Vint Cerf, one of the founders of technology that made sending and receiving data on the Internet possible, is working with NASA's Jet Propulsion Lab on a project to create the Interplanetary Internet.[57] In this video excerpt from a talk he gave at Google that the QR tag will take you to, Cerf explains how it is possible to connect planets to the Internet using satellites that orbit them. If all nations can adopt the networking standard he and NASA are proposing, Cerf sees the possibility of an interplanetary network backbone coming to life in a few decades.

Vint Cerf video on the Interplanetary Internet

This quick survey of communicants that are starting to

become connected to the Internet of Things, such as seas and satellites, captures a small sampling of projects in the works to create a global brain. Any one project could provide enough content for a book of its own. My purpose for sharing these is to give you a sense of the breadth of what is already happening, and as the pace of development becomes more rapid, it is worth considering how our lives may change in fundamental ways.

Gaining perspective of our place in the universe

We, as human beings, are but one among many communicants that already occupy our universe. The difference now is we are gaining the potential for hearing what the others have to say. Being part of a universal network has one more potential impact: it can shift our perspective about our relative place in the universe in nontrivial ways.

When I was sharing an early draft of the manuscript for this book with a friend with a good critical eye, she pointed out that I might want to think of a word other than "communicants" to describe all the elements that are starting to become connected to the Internet. "As someone who grew up Catholic," she pointed out, "a 'communicant' is what we call a person who partakes in Communion." I considered what she said, and in the end decided "communicant" is an apt term after all.

The concept of a "communicant" actually predates Christianity. The English word comes from Greek *koinonia,* whose meaning encompassed the notion of a sharing, generous, even virtuous mode of interaction with and among community that benefits the whole. Plato, in his dialogue *Gorgias,* even includes the gods as part of this communion.[58]

In contemporary language, you might say the very fact

that we are intimately connected in community kicks off potential positive feedback loops when engaged with a generous spirit. The open collaboration and sharing found among DIY communities today is a perfect example of the virtuous spiral spawned when people engage in community with a spirit of generosity: software gets improved more quickly than it would in a commercial environment; creativity and innovation flourish; and, in general, people are happier because they are doing what they want.

There is a kind of power generated through openness that is absent when our interactions are based in commerce. Many attendees of the Burning Man festival—with its "gifting community" ethos rather than one based on exchange—come away as I did, utterly surprised and delighted by how well a community governed by a spirit of giving works. The sheer amount of creative expression given freely, without any expectation of return, is astounding to even the most jaded participant. Both the open source DIY movement and Burning Man are living examples of communicants participating in *koinonia*.

I believe there is a way in which it makes sense to call the *nonhuman* entities joining the social network "communicants" participating in *koinonia* as well, without going so far as to anthropomorphize them. In a sense, everything from seas to cities and the technology they house has its own, unique contribution to make to the community driven by its own internal logic. We may build cities, but cities also shape us by prescribing certain behaviors to navigate their roadways, interact with their buildings and operate their machines. We cannot sail the sea without awareness of its waves and tides, which influences the size and shape of the boats we build and how we choose to get safely from shore to shore. In other words, we as human beings are but one among many communicants participating in this global community, but until now most of us did not have any way to make sense of how this might be true. With the advent of the Internet of Things, those connections will become increasingly tangible and may start to make sense. More importantly, having a tangible understanding of the contribution nonhuman entities make to our ecosystem actually gives us greater power to create the world we want. The question is, how?

The paradox of ego and agency

You might understand how making the feedback loops between ourselves and all that surrounds us tangible can help us comprehend how we shape and are shaped by our environment. What you might not have considered is its potential to shift our perspective from one where human beings are at the top of the food chain and run the Earth to one where we are simply one set of nodes in this massively networked world. Just as Copernicus sparked a scientific revolution by raising awareness that the Earth isn't the center of our universe, connecting our environment into a universal communication system can spark a revolution by raising awareness that we human beings are not the center of the universe.

The more accurately you understand your place in the universal ecosystem, the more power you have to affect change within it. This is the paradox of ego and agency. Too little ego diminishes your power to act effectively, as does too much ego. There is a "just right"-ness that can only be embodied through a shift in awareness. You can see a mundane example of this dynamic play out in childhood.

If you are a parent or have spent a lot of time around children, you well know that up until a certain point in a child's life it is all about him or her, and others are merely a source of food, shelter, and comfort. Until somewhere between five and eight, depending on the kid, the child doesn't really get that other people exist in their own right. And this is perhaps as it should be. It serves the child well to be self-centered and demanding to get their needs met until they have a strong foundation and are mature enough that they can take some responsibility for themselves. After that point, psychologically healthy children start to realize that they are one among many others with similar needs, drives, and desires. If they are smart, instead of demanding that others meet their needs, they will learn they have much *more* power to get what they want by understanding what the win-win might be for others. Conversely, if they remain egocentric they will learn that getting what they want will become harder and harder as the years go on.

We perhaps have played a similar role in the ecosystem. For a long time now we have been demanding what we want from the ecosystem without really being conscious that we are but one among many entities occupying this planet. And it has served us well, for the most part, for building a strong foundation and maturing to the point when we can take some responsibility for ourselves. Now that we can really get to know the other communicants in our midst, we can begin to understand what might be the win-win and discover better ways to get what we want from our environment. In a way, it is like moving from a commercial relationship with the world where we treat the environment as a resource from which we can demand payment, to one of *koinonia*—being in community governed by a spirit of generosity. If we can successfully make this transition, we will indeed be living in one smart universe. Making the transition, like transitioning out of childhood, calls for a shift in awareness.

Sparking a revolution

The potential for a shift in awareness of our place in the universe to spark a revolution was perhaps hinted at when the first astronauts reported their view of the Earth from space. Apollo 8 astronaut, Frank Borman, explained in a 1968 interview: "When you're finally up at the moon looking back on earth, all those differences and nationalistic traits are pretty well going to blend, and you're going to get a concept that maybe this really is one world and why the hell can't we learn to live together like decent people."[59] Jim Lovell, who rode both Apollo 8 and 13, described his experience: "We learned a lot about the Moon, but what we really learned was about the Earth. The fact that just from the distance of the Moon you can put your thumb up and you can hide the Earth behind your thumb. *Everything that you've ever known, your loved ones, your business, the problems of the Earth itself—all behind your thumb.* And how insignificant we really all are, but then how fortunate we are to have this body and to be able to enjoy loving here amongst the beauty of the Earth itself."[60]

In the case of these travels through outer space, the revolution was an individual one that manifested in their personal lives. Apollo 14 astronaut, Edgar Mitchell, also reported a profound shift in his consciousness when he saw the Earth from the Moon. In fact, he was so moved to share this newfound awareness that in 1973 he founded the Institute of Noetic Sciences (IONS) in California (www.noetic.org). IONS remains dedicated to research and education exploring consciousness, science, and spirituality to support the collective transformation of the planet nearly forty years later.

While the shift in awareness individual astronauts experienced after viewing Earth against the vast backdrop of space and the educational efforts that followed from some are important, they have had relatively little impact on the world as a whole. The reality is that the majority of us, myself included, are not that moved by the perspective of others unless that perspective connects with our *own* experience. Similarly, academic research or warnings about resource limitations only resonate with us when they become part of our experience. Chances are most of us will not be moved by what others say and will go about our daily lives as long as things seem to be rolling along as usual.

Until the early 1990s, I included myself among those content to live my day-to-day life without putting much thought into the ecosystem of which I was part. To be perfectly honest, I lumped all talk of ecological awareness and consciousness-raising into the same bucket of "things not that relevant to my life"—especially since I had grown up in the Bay Area of the 1970s and '80s, surrounded by what was often alarmist rhetoric from ecological activists and New Age idealists certain that the Harmonic Convergence or some such cosmic event would usher in cataclysmic environmental change. One ordinary weekday evening in 1991, sitting in my North Beach living room watching nothing particularly memorable on television, a startling realization hit me like the proverbial bolt from the blue. In a flash, I understood what those astronauts, ecologists, and New Agers might have been talking about. Like a deaf person who can all of a sudden hear, their words literally made sense. In that moment, I understood the sense in which we are already living in one smart universe. I got the substance of that awareness via a shift in consciousness rather than from something I learned from mundane experience.

I soon realized, however, that such flashes of insight are likely rare. So, I suppose much like the astronaut, Edgar Mitchell, who went on to found IONS, I started down my own path of considering how to connect what I now knew to be true with the experience of a greater number of people.

Fast forwarding to this moment, after nearly twenty years of following intuitive threads that I am now using to weave the fabric of this book, we are now beginning to be able to make the ecosystem a tangible part of our experience. One of the greatest realizations that I hope will come from us being part of this wired planet is that of our relative place in the universe. Participating in a tangible relationship with all the entities on our planet and beyond may help resolve the paradox of ego and agency as we gain a more accurate understanding of our place in the universe. And in that understanding, we actually gain more power to get what we want.

We need not wait until that transition is complete, though. We already have formidable tools with which to create something new—whether that is a new definition of health, a more fulfilling worklife, a better system for local governance, or a living environment that is more responsive to our needs. Moreover, we have started to build an infrastructure that allows each of us to become more tangibly aware of other communicants in our midst. Soon, I hope, each of us won't have to rely on a bolt from the blue or a trip into outer space to get a more accurate perspective of our place in the universe. As I wrote at the outset, social media and technology give you more power than ever before to create the world you want. And so does everyone and everything else. The tangible connections facilitated by social media and technology are giving each of us this power. Therein lies what makes *this* a unique time in human history to create the world you want.

Holding Ends and Beginnings Together

We are poised with one foot planted on the outer edge of a new world while the other stands in the realities of the old. Each of us is called to hold the ends and beginnings together. Imagination is the key.

The scenarios shared throughout this book perhaps more accurately describe an end rather than a beginning. The convergence of social media and technology sets the conditions that are transforming economic and ecological frameworks on a fundamental level. Not one of us knows yet what sort of new paradigms will emerge to replace the existing ones. All we do know is the gap between what we imagine and what becomes real is beginning to narrow.

As described in real-world examples throughout this book, the gap is narrowing quickly. We see it closing with the advent of technologies that can almost instantaneously augment or alter our perception of reality. What is real and what is virtual become less clearly distinguishable, so reality itself becomes more instantly malleable. As devices that connect us with technology become less obtrusive, the lines between biology and technology begin to vanish, friction in the feedback loop goes away and communication between the two approaches simultaneity. As we are networked together with the nonhuman world, instant access to the collective knowledge of the planet becomes easier and more expansive. And as personal manufacturing devices become more sophisticated and less expensive, instant production of goods becomes increasingly accessible. In a world where borders can be fluid and alliances can form among people from anywhere on the planet and disband when they are no longer needed, communities become evanescent. As the gap narrows, we reach a kind of atemporal equilibrium where the notion of "future" paradoxically becomes an anachronism.

When and if we reach a point where everything is changing all the time and in unpredictable ways due to the collective ability to manipulate reality in real-time and feed that real-time data into this massively networked world, planning your way toward a comfortable future based on what has worked before becomes increasingly elusive. You may soon be left no option but to take responsibility for imagining the world you would like to create based on whatever realities are currently shaping your experience. The good news is you can start today.

Allison and James, you may recall from earlier, assumed they were doing all the right things to secure a comfortable future only to find their assumptions fracturing as the economy unraveled beneath their feet. Their story, you may be glad to hear, ends on a happy note.

Allison took to heart my suggestion to imagine the world

she would like to create. She started by letting go of her assumptions about logical next steps by becoming conscious of the stories surrounding those assumptions—that climbing the corporate career ladder, investing in real estate and saving for retirement are viable paths to a secure future. She imagined the life she would like to create for her own fulfillment and the happiness of her family—one where she had a flexible schedule, wasn't dependent on a single employer and one where she earned more money. She looked at all the raw materials she had to work with—her expertise, unique talents, the need in the marketplace and her network of connections—and imagined how these might come together to create a new career. With a clear vision of what she wanted, Allison created a new career for herself as an independent consultant. She is now able to take an active role in shaping her worklife while earning an even better living than she did before. She made this transition from imagination to securing her first client within only a few months.

Allison's story exemplifies the practical power of imagination. Imagination is not a passive faculty. The imaginative faculty can help you tune in to resources and ways to use them you may not have considered. The first critical step is letting go of old stories and assumptions about what is and is not possible. The next step, starting from a clean slate of unlimited potential realities, is imagining what you would like to create. This may be as simple as creating a solution for removing litter from your street or it may be as grand as eradicating world hunger. For most of us, letting go of old stories and assumptions and engaging the imagination will take a leap of faith. Like that dog who never ventures beyond the six-foot circle that defined his territory while he remained leashed, we may need someone to point out the leash is gone and the circle is now as big as the universe. Scary to consider at first, perhaps, but when the territory within which your life is defined ends up directly in the path of a tornado, best have done a little scouting for new territory in advance.

Allison and James now know they can keep one foot grounded in the current world while they gain a foothold in the world they want to move toward. After taking the leap of faith, they are gaining more confidence that by taking responsibility for the world they imagine, they can realize the future they want. They do so with the faith that their son Malcolm will grow up in a world they helped hold together—a massively networked world where each of us accurately understands our place in the universe and our responsibility to create the world we want, individually and collectively.

Notes

Introduction - Welcome to the Massively Networked World

1. Jessica Reinis, "Children's 'Future Requests' for Computers and the Internet," (Latitude 42's Open Innovation Series, July 2010) http://www.life-connected.com/cms/wp-content/uploads/2010/07/Latitude-Research-42-KidsTech-Study-Summary.pdf.

2. IMS Research, "Internet Connected Devices About to Pass the 5 Billion Milestone," August 19, 2010 press release.

Chapter 1 - Your Creative Reality

3. What might have Einstein meant by saying imagination is more important than knowledge? Here is one idea from Philosophy 101 to ponder if you haven't before: Logic, knowledge, and what most people consider "rational" thought make sense *only* within a frame of reference. If this doesn't make sense to you, ask yourself: If everything we all see were the same shade of yellow, how would you know what "color" is? If all things were the same shade of yellow, we would have no frame of reference within which to get what "color" might mean. However, once both yellow and, say, blue are perceptible, we can imagine a "color" framework for making logical sense of how yellow is, and is not, like blue—and *vice versa*. The category of color organizes your perception of a type of sensory experience—in this

case the qualities of yellowness and blueness—so that you can formalize this experience into a piece of knowledge. This is one way imagination can be said to be more important than knowledge. Imagination can hold the perception of both yellowness and blueness—almost like puzzle pieces—and put together a framework you can call "color" that makes knowledge of yellow and blue—as well as other colors—possible.

4. Studio 360, "Stewart Copeland & Zee Avi," September 2, 2010, http://www.studio360.org/episodes/2010/09/03.

Chapter 3 - Your Massively Networked Self

5. Gary Wolf, telephone interview with author, November 11, 2010.
6. Michelle Bryner, "Smart clothing responds to wearer's emotions," *Tech News Daily,* June 8, 2010, http://www.technewsdaily.com/smart-clothing-responds-to-wearers-emotions-0669/.
7. Department of NanoEngineering at the University of California, San Diego Jacobs School of Engineering, "NanoEngineers Print and Test Chemical Sensors on Elastic Waistbands of Underwear," June 16, 2010 press release, http://www.jacobsschool.ucsd.edu/news/news_releases/release.sfe?id=958.
8. Lauren Gravitz, "Glaucoma test in a contact lens," *Technology Review,* March 31, 2010, http://www.technologyreview.com/business/24931/.
9. University of Western Ontario, "Nanocomposites could change diabetes treatment," *Western News,* December 16, 2009, http://communications.uwo.ca/com/western_news/stories/nanocomposites_could_change_diabetes_treatment_20091216445482/.

10. Jon Weiner, "Caltech-led Team Provides Proof in Humans of RNA Interference Using Targeted Nanoparticles," *California Institute of Technology News*, March 21, 2010, http://media.caltech.edu/press_releases/13334.

11. Katie Moisse, "Plastic Fantastic: Synthetic Antibodies Recognize and Remove Toxins in Mice," *Scientific American*, June 9, 2010, http://www.scientificamerican.com/blog/post.cfm?id=plastic-fantastic-synthetic-antibod-2010-06-09.

12. Lauren Gravitz, "Old livers made new again," *Technology Review*, June 14, 2010, http://www.technologyreview.com/biomedicine/25538/?a=f.

13. "Clinical transplatation of a tissue-engineered airway," *The Lancet*, Volume 372, Issue 9655, pages 2023-2030, December 13, 2008, http://www.thelancet.com/journals/lancet/article/PIIS0140-6736%2808%2961598-6/fulltext.

14. Mitch Leslie, "Rats breathe with lab-grown lungs," *Science Now*, June 24, 2010, http://news.sciencemag.org/sciencenow/2010/06/rats-breathe-with-lab-grown-lung.html.

15. University of Missouri webpage: "Organ Printing: Understanding and employing multicellular self-assembly," http://organprint.missouri.edu/www/index.php.

Chapter 4 - Your Daily Life: Smarter and More Efficient

16. Larry Hardesty, "Gesture based computing on the cheap," *MIT News Office*, May 20, 2010, http://web.mit.edu/newsoffice/2010/gesture-computing-0520.html.

17. Kristina Grifantini, "Mobile phone mind control," *Technology Review*, March 31, 2010, http://www.technologyreview.com/blog/editors/24993/?a=f.

18. Chris Meyers, "Japan scientists create 3D images you can

touch," *Reuters*, September 16, 2009.

19. Massachusetts Institute of Technology webpage: "Cornucopia: Prototypes and Concept Designs for a Digital Gastronomy," http://web.media.mit.edu/~marcelo/cornucopia/.

20. Cornell Computational Synthesis Laboratory webpage: "Printing Food," http://ccsl.mae.cornell.edu/node/194.

21. Chris Harnick, "IKEA uses augmented reality to launch PS furniture collection," *Mobile Marketer*, November 26, 2009, http://www.mobilemarketer.com/cms/news/content/4729.html.

22. Ray Ban Virtual Mirror webpage: http://www.ray-ban.com/usa/science/virtual-mirror.

23. United States Post Office Priority Mail webpage: https://www.prioritymail.com/simulator.asp.

24. Massachusetts Institute of Technology webpage: "Precision Agriculture: Sustainable Farming In The Age Of Robotics," http://www.csail.mit.edu/csailspotlights/feature2.

25. Sweet Water Organics webpage: http://sweetwater-organic.com/blog/.

26. Wailin Wong, "Gap's Groupon pulls in $11 million," *Chicago Tribune*, online August 20, 2010, http://articles.chicagotribune.com/2010-08-20/business/sc-biz-0821-groupon-20100820_1_gender-and-zip-code-chicago-startup-coupon-site.

27. A series of articles on creating intelligent environments available at University of Washington publications webpage: http://ailab.wsu.edu/casas/pubs.html. for

28. A. Crandall and D. Cook, "Coping with multiple residents in a smart environment," *Journal of Ambient Intelligence and Smart Environments*, 2009.

29. Auger-Loizeu, *HappyLife* webpage: http://www.auger-

loizeau.com/index.php?id=23.

Chapter 5 - Rethinking Your Worklife

30. Kim-Mai Cutler, "300 Million Tweets Reveal the Afternoon at Work Is the Unhappiest Time of Day," *New York Times*, July 23, 2010, http://www.nytimes.com/external/venturebeat/2010/07/23/23venturebeat-300-million-tweets-reveal-the-afternoon-at-w-93240.html.

31. Christopher Peterson, "Money and happiness," *Psychology Today*, June 6, 2008, http://www.psychologytoday.com/blog/the-good-life/200806/money-and-happiness.

32. Jonah Lehrer, "Why money makes you unhappy," *Wired*, July 21, 2010, http://www.wired.com/wiredscience/2010/07/happiness-and-money-2/.

33. Bernard Lietaer, "Call for our times," http://www.lietaer.com/2010/09/callofourtimes/.

Chapter 6 - The DIY Community: Creating the World You Want

34. Mallory Simon, "Island DIY: Kauai residents don't wait for state to repair road," *CNN*, April 9, 2009, h t t p : / / www.cnn.com/2009/US/04/09/hawaii.volunteers.repair/index.html.

35. BQF Innovation website: "How to get a whole country brainstorming," http://www.bqf.org.uk/innovation/2009/04/14/how-to-get-a-whole-country-brainstorming/.
Let's Do it Estonia! website: http://www.minueesti.ee/?lng=en&leht=88,101.

36. John Geraci, telephone interview with author, November 3, 2010.

37. Open311 website: http://open311.org/learn/.

38. Dina Cappeillo, "Mark Bly, BP oil spill lead investigator,

admits limitations with internal probe," *Huffington Post*, September 26, 2010, http://www.huffingtonpost.com/ 2010/09/27/mark-bly-bp-oil-spill-inv_n_740050.html.

39. BP *"Deepwater Horizon* Accident Investigation Report," September 8, 2010, http://www.bp.com/liveassets/bp_internet/globalbp/globalbp_uk_english/incident_response/ STAGING/local_assets/downloads_pdfs/Deepwater_ Horizon_Accident_Investigation_Report.pdf.

40. Planet Money Podcast, "Good News from Haiti," *NPR*, July 2, 2010, http://www.npr.org/blogs/money/2010/07/02/ 128270251/the-friday-podcast-haiti.

41. Jeffery Tumlin, telephone interview with author, October 25, 2010.

42. Ties Van Der Werff, "Social networking with plants," June 28, 2010, *Next Nature*, http://www.nextnature.net/2010/06/ social-networking-with-plants/.

43. Charles Crain, "Army mechanic's garage tinkering yields 18-foot mecha exoskeleton," *PopSci*, June 8, 2009, http:// www.popsci.com/scitech/article/2009-05/man-machine/. Carlos Owens' website: http://neogentronyx.com/index. php.

44. Chris Pollard, "Spiderlad," *The Sun UK*, June 29, 2010, http://www.thesun.co.uk/sol/homepage/news/3033607/ Hibiki-Kono-climbs-a-brick-wall-after-turning-himselfinto-Spiderman.html.

45. Bre Pettis, email interview with author, November 4, 2010.

46. Priya Ganapati, "Frugal hobbyists put satellites into orbit with 'TubeSat,'" *Wired.co.uk*, July 21, 2010, http://www. wired.co.uk/news/archive/2010-07/21/satellite-tubesathobbyists.

47. SXM Project, "Scanning tunneling microscope construc-

tion kit," http://sxm4.uni-muenster.de/stm-en/.
48. Zain Jaffer, "Eri Gentry's garage biotech revolution," *Wired.co.uk*, July 15, 2010, http://www.wired.co.uk/news/archive/2010-07/15/eri-gentry-garage-biotech-revolution.

Chapter 7 - Living in One Smart Universe

49. Rice Wisdom: K.P.M. Basheer, "The shrinking rice fields of Pokkali," http://ricewisdom.org/shrinking-fields-of-pokkali.html.
50. India Environmental Portal: "In troubled waters," December 14, 2005, http://www.indiaenvironmentportal.org.in/node/19822.
51. Miguel Bustillo, "Walmart to put radio tags on clothes," *Wall Street Journal Online*, July 23, 2010, http://online.wsj.com/article/SB1000142405274870442130457538321306 1198090.html?mod=WSJ_hps_MIDDLEForthNews.
52. Christopher Mims, "Cape Cod is tweeting, thanks to the Internet of Things," *Technology Review*, June 11, 2010, http://www.technologyreview.com/blog/mimssbits/25315/.
53. "Cisco wires 'city in a box' for fast-growing Asia," *News Observer*, June 8, 2010, http://www.newsobserver.com/2010/06/08/v-print/520176/cisco-wires-city-in-a-box-for.html.
54. Lisa Fletcher and Andrea Beaumont, "Boulder, Colo.: America's first 'Smart Grid City,'" *ABC News*, November 15, 2008, http://abcnews.go.com/GMA/SmartHome/story?id=6255279&page=1.
55. HP Information and Quantum Systems Lab webpage: http://www.hpl.hp.com/research/quantum_systems/.
56. "The 50 best inventions of 2009: Planetary Skin," *Time*, http://www.time.com/time/specials/packages/article/0,28804,1934027_1934003_1933962,00.html.

57. "Interplanetary Internet," Wikipedia, http:/en.wikipedia. org/wiki/Interplanetary_Internet.

58. Plato, *Gorgias* (New Jersey: Prentice-Hall, Inc., 1952), 83-84. Section 508a reads "...and gods and men are held together by communion and friendship, by orderliness, temperance, and justice; and that is the reason, my friend, why they call the whole of this world by the name of order."

59. Space Quotations Looking Back at Earth webpage quotes Frank Borman, Apollo 8 astronaut, from *Newsweek*, December 23, 1968, http://www.spacequotations.com/ earth.html.

60. Space Quotations Looking Back at Earth webpage quotes Jim Lovell, Apollo 8 & 13 astronaut, interview for the 2007 movie, *In the Shadow of the Moon*, http://www.space-quotations.com/earth.html.

Index

Acknowledgments

Massively Networked owes its existence to the generous support of many friends, family and colleagues. Many thanks goes to Ingrid Stabb and Carolyn Foss, who provided valuable feedback from the first chapter outlines to beginning drafts of the manuscript. Their early support made facing a blank page much easier, and their encouragement inspired me to keep writing. Thanks to my developmental editor, Erin Reese, for reordering chapters of the first draft, taking some out altogether and expertly editing the rest. With her guidance, *Massively Networked* is no doubt a much more enjoyable read. Thanks as well to colleagues Risa Dimacali, Betsy Flanagan and Korina Park, for contributing their keen perspectives on marketing and media over months of discussion.

Finally, thank you to all who publicly supported the Massively Networked IndieGoGo campaign with a contribution of $50 or more: Dorit Adams Richardson of Purple Rose Software, Gary Angel of Semphonic, Renee Anker, Gregg Butensky of madnomad.com, Risa Dimacali, Betsy Flanagan of Startup Studio, Carolyn Foss of Forté Strategic Business Consulting, Linda Leong, Korina Park of KD Idea Inc. Online Business Consultants, Erin Reese of Travel and Soul Media, Becky Snyder, Web and Database Marketing Analyst, Ingrid Stabb, Co-Author of The Career Within You and Paul Young. Your generous financial support helped take this book from concept to reality.

About the Author

Pamela Lund is a practical visionary and keen observer of life: whether from behind the scenes of NYC punk clubs and raves of the early '80s, perched on the saddle of a motorcycle traveling solo through rural northeast Thailand, or from the perspective of startup entrepreneur. She holds a B.A. in Philosophy from UCLA and a Masters in Theological Studies from Harvard. Pamela divides her time between rural Montana and San Francisco, where she works as an interactive marketing consultant, trend-spots, kayaks and occasionally likes to stir things up.

CPSIA information can be obtained at www.ICGtesting.com
Printed in the USA
LVOW110842050312

271633LV00002B/10/P